AA003383

Advances in Sustainability

Topical Conference at the 2012 AIChE Spring Meeting and 8th Global Congress on Process Safety

Houston, Texas, USA
1-5 April 2012

ISBN: 978-1-62276-563-8

Printed from e-media with permission by:

Curran Associates, Inc.
57 Morehouse Lane
Red Hook, NY 12571

Some format issues inherent in the e-media version may also appear in this print version.

Printed by Curran Associates, Inc. (2012)

For permission requests, please contact AIChE
at the address below.

AIChE
3 Park Avenue
New York, NY 10016-5991

Phone: (203) 702-7660
Fax: (203) 775-5177

www.aiche.org

Additional copies of this publication are available from:

Curran Associates, Inc.
57 Morehouse Lane
Red Hook, NY 12571 USA
Phone: 845-758-0400
Fax: 845-758-2634
Email: curran@proceedings.com
Web: www.proceedings.com

TABLE OF CONTENTS

BIOMASS TO OIL: FAST PYROLYSIS AND SUBCRITICAL HYDROTHERMAL LIQUEFACTION

Nicolas Doassans-Carrère, Laboratoire M2P2, Aix-en-Provence, France
Guillain Mauviel, LRGP, Nancy, France
Jean-Henry Ferrasse, Laboratoire M2P2, Aix-en-Provence, France
Jacques Lédé, LRGP, Nancy, France
Olivier Boutin, CEA, Marcoule, France

Abstract

The present paper deals with the comparison of two biomass-to-oil processes: fast pyrolysis and subcritical hydrothermal liquefaction. Using the same biomass (beech sawdust), fast pyrolysis was led thanks to a cyclone reactor (wall temperature between 870 and 1040 K) and subcritical hydrothermal liquefaction thanks to a 150-ml-batch-reactor (temperature between 420 and 600 K). Mass balances and phases analysis (ultimate analysis, higher heating value (HHV), Karl-Fischer, gas chromatographies, H^1 NMR) allow the comparison of both processes and the characterization of pyro-oils (heavy oils, light oils and aerosols), liq-oils (heavy oils and water soluble organics), solids and gas.

For both processes, there is a temperature for which a high conversion of biomass is reached with a minimum of gas mass yield. Up to 62.6 wt% of pyro-oil and 47.0 wt% of liq-oil, on dry basis, were obtained in this work, with oxygen contents and chemical compositions different from one to the other.

Introduction

The global energy demand is continuously increasing, due to the growth of both population and industrialization. The supply of CO_2-neutral energies has to become more important in the future, and bio-fuels from biomass may be a part of those energies.

Different processes have been developed to produce bio-oils from different biomass, among them fast pyrolysis and subcritical hydrothermal liquefaction are studied in this work. Fast pyrolysis can produce up to75%wt of bio-oil (pyro-oil), composed typically of 56%wt of C, 38%wt of O and 6%wt of H [1]. Subcritical hydrothermal liquefaction consists in thermal decomposition of biomass in water in high pressure and temperature condition but below its critical point (647 K and 22.1 MPa). Subcritical water is mainly used today for wet air oxidation [2]. Using subcritical hydrothermal liquefaction, 35%wt of bio-oil (liq-oil) can be reached. This liq-oil, less oxygenated than pyro-oil, is composed typically of 74%wt of C, 17%wt of O and 9%wt of H [1]. Those two types of bio-oils must be upgraded to obtain bio-diesel, thanks to catalytic hydrodeoxygenation (HDO) which reduces the proportion of oxygen.

The aim of this paper is a direct comparison of fast pyrolysis and subcritical hydrothermal liquefaction for the production of bio-oil. This comparison is based on mass balances and products characterization: bio-oils, solids and gas.

Experimental apparatus

In order to compare those processes, experimentations were led with the same biomass, beech sawdust.

Fast pyrolysis

Fast pyrolysis is achieved thanks to the cyclone reactor [3] which is a continuous reactor with an oil recovery system separating three fractions of pyro-oil (Figure 1). Up-stream of the cyclone, the installation is composed of the biomass (tank and screw conveyor and nitrogen (inert carrying gas) feeds, with a mass flowrate ratio between nitrogen and biomass equal to 7.8. Particles, carried by nitrogen, are pyrolysed along the cyclone wall which is heated by induction. A by-pass system allows preheating and equilibrium set-up before biomass feeding. Thanks to the separator properties of the cyclone, solids are isolated from gaseous products and recovered in the solids collector. Gaseous products, evacuating through the upper part of the cyclone, pass through three water cooled heat exchangers in series to recover the first fraction of oil in a collector, "heavy oils" (HO). Then the gas flows through refrigerated coils inside which the "light oils" (LO) fraction is recovered. In order to recover the last fraction, "aerosols" (AE), both electrostatic and membrane filters are used in series. At the exit, clean gases are finally sampled and analyzed by micro-GC.

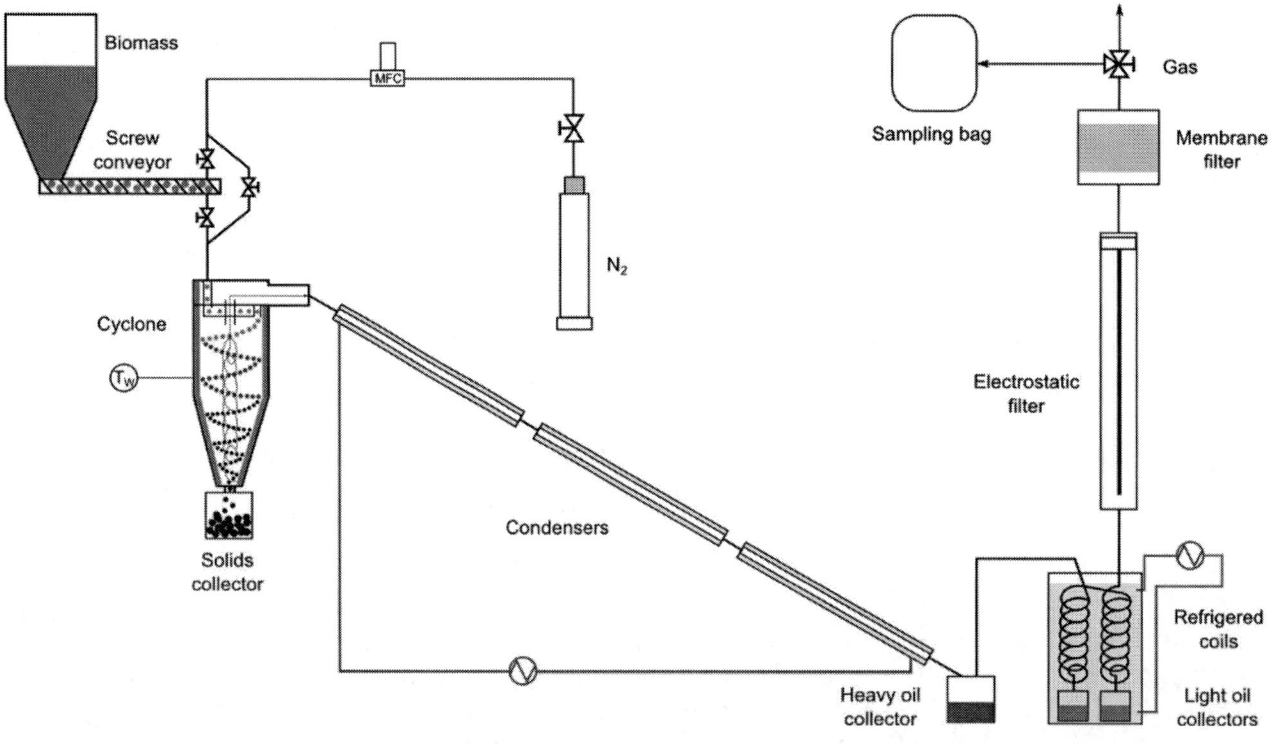

Figure 1. Fast pyrolysis cyclone reactor installation scheme

Subcritical hydrothermal liquefaction

Subcritical hydrothermal liquefaction is achieved in a 150 ml autoclave where biomass (6 g) is introduced prior to the catch of the reactor with the locking straps (Figure 2). Thanks to a volumetric pump, nitrogen or distilled water can be introduced into the autoclave at a desired pressure. After inerting the reactor with nitrogen and setting the initial working pressure (1 MPa), 100 ml of distilled water are added into the reactor, heated by an electric resistance. Once achieved, the working temperature is maintained at the experiment duration time, then the reactor is cooled down by a refrigerating system using air and then water. Once the reactor at the ambient temperature, gases are sampled and analyzed by GC-TCD. The reactor is opened in order to collect the products (liquids and solids). Based on [4], pure acetone is used to rinse the reactor and wash solids. After filtration, three phases are obtained: solids, aqueous phase and organic phase. Solids are placed into a drying oven. Evaporation at reduced pressure is used to remove acetone from the organic phase and water from the aqueous phase, in order to get respectively two oil fractions, "heavy oils" (HO) and "water soluble organics" (WSO).

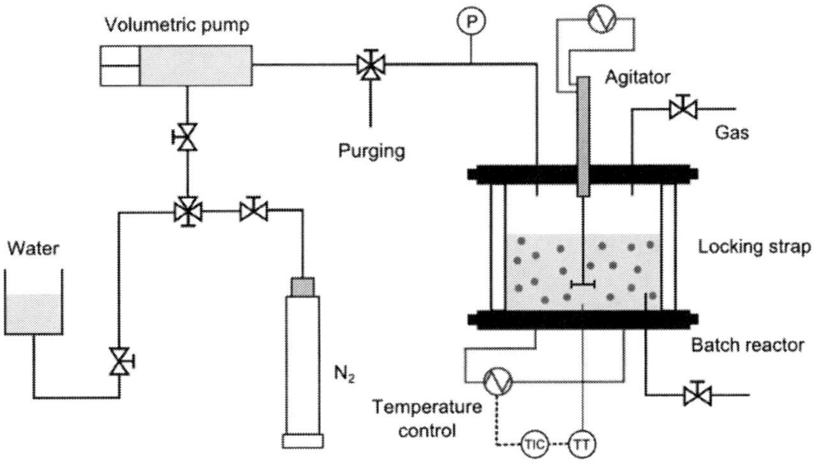

Figure 2. Subcritical hydrothermal liquefaction installation scheme

Results and discussion

Maximum of bio-oil mass yield

For both installations, mass balances are based upon mass measurement of solids and bio-oils, and evaluation of the gas composition. Thanks to the known amount of nitrogen used in those installations (mass flowrate for fast pyrolysis and initial pressure for subcritical hydrothermal liquefaction), the mass of produced gas can be determined from gas concentrations in the gaseous phase. Mass yield are defined from dried biomass, with a biomass moisture of 9 wt.%. It has to be noticed that for fast pyrolysis, after data reconciliation based upon atom balances, mass balances close at 100 % (the initial overall mass balances closures range between 83 and 100 %). No data reconciliation could be made for subcritical hydrothermal liquefaction, and the overall mass balances closures range between 64 and 83 %. This misclosure is often explained by a water production [5].

Figure 3 shows influence of temperature on mass balances for fast pyrolysis and subcritical hydrothermal liquefaction. For both processes, there is a temperature for which a maximum of bio-oil mass yield is reached. For fast pyrolysis, at 970 K, 62.6 wt% of pyro-oils are obtained with 25.7 wt% of gas and 11.7 wt% of solids. For subcritical hydrothermal liquefaction, at 573 K, 47.0 wt% of liq-oils are obtained with 5.5 wt% of gas and 17.8 wt% of solids. This maximum correspond to a high conversion of biomass coupled with a low gas mass yield. The partial conversion of biomass is revealed by ultimate analysis of solids recovered. In fast pyrolysis, the composition of solids recovered at 870 K ($CH_{1.26}O_{0.57}$) is closer to initial biomass one ($CH_{1.48}O_{0.72}$) than chars ones obtained at 1040 K

($CH_{0.24}O_{0.17}$). In subcritical hydrothermal liquefaction, until 528 K, the composition of solids ($CH_{1.50}O_{0.68}$) is similar to the biomass one, whereas from 573 K, solids recovered are closer to chars ($CH_{0.75}O_{0.24}$).

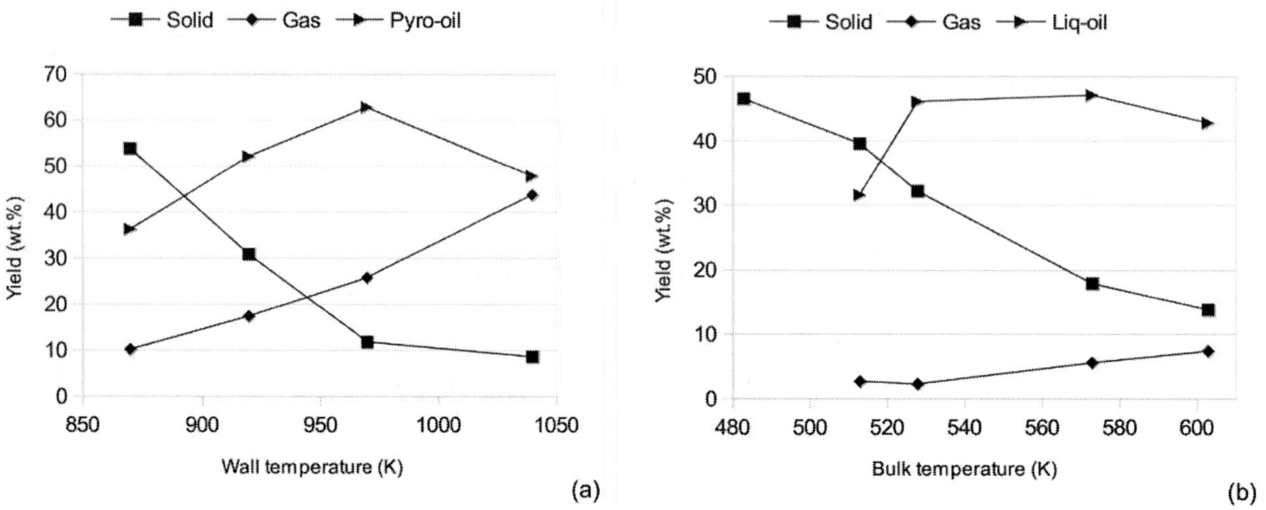

Figure 3. Mass yields evolutions according to temperature for (a) fast pyrolysis and (b) subcritical hydrothermal liquefaction

Products characterization

Apart from ultimate analysis made on solids, which have revealed partial biomass conversion, characterization of gas and bio-oils allows the direct comparison between both fast pyrolysis and subcritical hydrothermal liquefaction.

Gas from fast pyrolysis is a mixture of CO, CO_2, H_2, CH_4, C_2H_4 and others C_2-C_3, with a majority of CO, CO_2 and CH_4 (> 91 wt.%). At the opposite, gas from subcritical hydrothermal liquefaction is mainly composed by CO_2 (> 99 wt.%), with a maximum of 1 wt.% of H_2 at 603 K.

Pyro-oil is the most oxygenated bio-oils with composition of heavy oil, light oil and aerosols of, respectively, $CH_{3.34}O_{1.73}$, $CH_{4.08}O_{1.98}$ and $CH_{1.32}O_{0.46}$. Composition of liq-oil fractions (heavy oil and water soluble organics) are respectively $CH_{1.11}O_{0.43}$ and $CH_{1.65}O_{0.78}$. This difference of oxygen content is partly explained by the water content of pyro-oil. In fact, contrary to liq-oil free of water (consequence of evaporation steps), 40 wt.%, 36.5 wt.% and 6.5 wt.% of, respectively, heavy oil, light oil and aerosols is water.

Gas chromatographies and H^1 NMR confirm this oxygen content difference. Using the same analytical apparatus, a direct comparison is possible between pyro-oil and liq-oil. Those bio-oils have some compounds in common, like phenol, guaiacol or acetic acid, but major compounds are different. For example, aldehydes or carbohydrates like levoglusan, major pyro-oil oxygenated compounds, are absent in liq-oil. At the opposite, hydrogen from alkanes or aromatics, are in higher proportion in liq-oil, indicating longer-chain compounds (and therefore less oxygenated) or higher quantity of compounds from those families.

Finally, ultimate analysis and HHV-measurement, made on solids and fractions of pyro-oils and liq-oils, confirm the use of a correlation Dulong-like between both of those properties [6]. Indeed, liq-oils have higher HHV than pyro-oils, but with a lower mass yield. Energetic balances are close to mass balances with up to 59.5 % and 51.5 % of initial biomass HHV are recovered, respectively, in pyro-oils and liq-oils.

Conclusions

Fast pyrolysis and direct liquefaction of beech sawdust has been studied in the production of bio-oils, through mass balances and products characterization. For both processes, a maximum of bio-oils mass yield is observed which corresponds to a high conversion of biomass coupled with a minimum gas mass yield.

The higher oxygen content of pyro-oils than liq-oils, revealed by ultimate analysis, is explained by the water content and different compositions, shown by GC-MS/FID and H1 NMR analysis.

For future work, those results are used as a basis for energy analysis which must be achieved to ensure the viability of the two processes for the production of bio-diesel, including HDO step. This step is, in particular, simulated thanks to bio-oils analysis like oxygen content and compounds families.

References

1. Huber, G. W., Iborra, S., and Corma, A. (2006). Synthesis of transportation fuels from biomass : Chemistry, catalysts, and engineering. *Chemical Reviews*, 106(9) :4044–4098.
2. Lefèvre, S., Boutin, O., Ferrasse, JH., Malleret, L., Faucherand, R. and Viand, A. (2011). Thermodynamic and kinetic study of phenol degradation by a non-catalytic wet air oxidation process. *Chemosphere*, 84 (9) :1208-15.
3. Lédé, J., Broust, F., Ndiaye, F.-T. and Ferrer, M. (2007). Properties of bio-oils produced by biomass fastpyrolysis in a cyclone reactor. *Fuel, 86*, 1800-1810
4. Sun, P., Heng, M., Sun, S.-H., and Chen, J. (2011). Analysis of liquid and solid products from liquefaction ofpaulownia in hot-compressed water. *Energy Conversion and Management*, 52(2) :924 – 933.
5. Cheng, S. D'cruz, I., Wang, M., Leitch, M., and Xu, CC (2010). Highly efficient liquefaction of woody biomass in hot-compressed alcolhol-water co-solvents. *Energy & Fuels*, 24(9):4659-67
6. Channiwala, S. and Parikh, P. (2002). A unified correlation for estimating hhv of solid, liquid and gaseous fuels. *Fuel*, 81(8):1051

(62b) Pyrolysis Oil and Its Applications

Rajeswar R. Gattupalli, Process Design Development, UOP LLC, A Honeywell Company, Des Plaines, IL, Stanley Frey, Renewable Energy & Chemicals, UOP LLC, Des Plaines, IL and Lance Baird, UOP LLC, A Honeywell Company, Des Plaines, IL

Fuels obtained from biomasses are becoming a valid alternative to the use of fossil fuels, especially in the light of stringent environmental constraints.Fast pyrolysis of lignocellulosic biomass produces a renewable liquid fuel called pyrolysis oil that has a variety of applications as a potential replacement for fossil fuels.One aspect of pyrolysis oil that has been addressed to promote its broader use is improving the stability of pyrolysis oil with respect to viscosity increase over time, making long range transportation and storage possible. Other properties of pyrolysis oil present challenges during its use in standard equipment constructed for petroleum derived fuels. Significant effort has been spent on research and development to improve the stability of pyrolysis oil and for its use in burners and other energy applications.

This presentation is divided into two parts - in the first part, we will give a brief background on pyrolysis oil and its stability concerns, the second part will focus on the rapid thermal processing or RTP process for fast thermal conversion of wood and/or other biomass to high yields of liquid fuels and the various applications of stabilized RTP green fuel.

Catalytic performance of Mn-promoted Ni/MgO catalysts in CO_2 reforming of toluene used as model compound of tar produced from biomass: Influence of preparation methods

Meng Kong, Jin-hua Fei*,Xiao-ming Zheng*

Institute of catalysis, Xixi Campus, Zhejiang University, Hangzhou, P. R. China, 310028

Abstract

Catalytic performance of Mn-promoted Ni/MgO catalysts prepared by different methods was investigated in CO_2 reforming of toluene. By XRD, H_2-TPR and TEM analysis and reforming tests, catalysts made by either conventional impregnation or precipitation methods in which $Mn(NO_3)_2 \cdot 4H_2O$ was used as the start precursor behaved similarly in catalytic performance, Mg_6MnO_8phase was present in all the calcined catalysts which caused worse activities in reforming tests. In contrast, by using a mechanic milling method in which MnO was initially added by physical milling with MgO as a support and then the calcined support was impregnated with nickel nitrate, the main crystalline phase of Mn species was α-Mn_2O_3 which worked as a support, after reduction, Ni can distribute well on both MnO and MgO, in this case, more Ni particles can be obtained because the interaction between Ni and MnO was less strong than that with MgO, therefore, a better performance than Ni/MgO and other catalysts can be yielded.

Keywords: Tar, Reforming, Ni/MgO, Mg_6MnO_8, α-Mn_2O_3

Introduction

Tar is an inevitable by-product during biomass gasification which can pollute environment, block process pipes and decrease calorific value.[1-3]Reforming of tar compounds to syngas is an efficient method to improve the overall energy recovery during gasification while minimizing environmental pollution.[2]

Many researchers have used steam as an oxidizing agent for reforming tar to produce H_2-rich syngas. However, the concentration of steam in the rough production gases from biomass gasification is much less than the concentration of CO_2 which can range from 10 to 30 vol. %.[3,4]Therefore, CO_2 could be a better choice than H_2O for reforming the tar to boost the effective conversion of carbon and minimize the release of CO_2 into the atmosphere by providing more syngas. In our study, toluene was selected as a tar model compound because it represented one of the major tar species 错误!未定义书签。, after comparing some supported catalysts, Ni/MgO was recognized as an efficient catalyst in CO_2 reforming of toluene, 80% of toluene was transformed to syngas at 600℃with a CO_2/toluene mole ratio of 14.[5]In order to improve the catalytic performance especially at a lower reaction temperature, experimental study and Co-Ni bimetallic catalysts have been investigated.

In this contribution, MnO was used as a promoter, because it showed good capacities in stabilizing Ni particles and eliminating coke in Ni/Al_2O_3catalyst for dry-reforming of methane.[6,7] However, the conventional impregnation and precipitation methods were not suitable in Ni-MnO-MgO catalyst system. Therefore, in this paper, the influence of preparation method of MnO-promoted Ni/MgO catalyst in CO_2 reforming of toluene was investigated, which was helpful for readers to understand the relationship between catalytic performance and catalysts' structure.

2. Experimental Section

2.1. Catalyst preparation

Ni/MgO, Ni/MnO and Ni-MnO-MgO catalysts were prepared by the following methods. In all catalysts the loading of Ni was controlled at 5% and MnO was 10% by weight, in Ni/MnO, the loading of Ni was also 5%, the rest part was MnO.

Ni/MgO was prepared by impregnating with Ni $(NO_3)_2 \cdot 6H_2O$ aqueous solution and MgO overnight, and then dried at 120℃ for 6h, finally calcined at 700℃ for 6h.

Ni-MnO/MgO-CI catalyst was made by a co-impregnation method. MgO was impregnated initially with Ni $(NO_3)_2 \cdot 6H_2O$ and 50 wt.% Mn $(NO_3)_2$ aqueous solution overnight, then followed the above steps.

Ni/MnO-MgO-SI catalyst was made by a sequential impregnation method. MnO/MgO-IM support was prepared first by the impregnation method as same as Ni/MgO, then the modified MnO/MgO support was impregnated with Ni $(NO_3)_2 \cdot 6H_2O$ solution, dried and calcined at the same conditions above.

Ni/MnO-MgO-PI catalyst was made by a precipitation-impregnation method. MnO/MgO-PR support was first prepared by the precipitation method in which50 wt.% Mn $(NO_3)_2$aqueous solution was used as precursor, and Na_2CO_3 was the precipitation agent, and the PH was controlled at 10. After dring and calcination, the modified support was impregnated with nickel nitrate solution as Ni/MnO-MgO-SI did.

Ni/MnO-MgO-MI catalyst was also made by an impregnation method. First, the support of MnO/MgO-MM was prepared by mechanic milling MnO_2 with MgO uniformly, and then calcined at 700℃ for 6h, this support was finially used to prepare Ni-based catalyst by an impregnation method as above.

Ni/MnO_2 catalyst was prepared by co-precipitation with aqueous solution of Ni $(NO_3)_2 \cdot 6H_2O$ and Mn $(NO_3)_2$ and Na_2CO_3 as precipitation agent. Ni/MnO_2 was not made by impregnation, because MnO_2 has a badly worse hydrophilic capacity and it is very hard to squeeze to a suitable size of 0.3-0.45mm..

2.2. Catalyst characterization

The crystalline phases contained in the samples were examined by powder X-ray diffraction system (Rigaku, D/MAX 2500/PC) using Cu Ka radiation (45 kV, 40 mA).

The TEM experiments were performed by using JEM-2010(JEOL Ltd.) TEM instrument operated at 200 kV. An X-ray analyzer for EDS is incorporated into the instrument for elemental analysis under STEM mode. The sample was first dispersed in ethanol and supported on Lacey-formvar carbon on a 200 mesh Cu grid before the TEM images were recorded.

2.3. Catalyst activity test

CO_2 reforming of toluene was carried out at atmospheric pressure in a tubular quartz fluidized-bed reactor having an inside diameter of 9.8 mm. The catalyst (0.5 ml, 0.3-0.45 mm) was first reduced at 700 ^{o}C in H_2 for 60 min and then tested for 450 min at 570 ^{o}C. For testing, liquid toluene was supplied at a rate of 0.3 ml/h byan HPLC pump (ASI, 501) into a constant temperature evaporator (120°C) where a gas mixture composed of 285 ml/min Ar and 14.5 ml/min CO_2 (atmospheric pressure) was introduced at the same time. The flow rates of CO_2 and Ar were measured by mass flowmeters (Brooks, 5850E).

The effluent gases were analyzed with an on-line gas chromatograph (Agilent, GC-6820) equipped with a TDX-101 column and a thermal conductivity detector (TCD) to determine the H_2, CO, and CO_2 concentrations (No CH_4 was detected). Simultaneously, toluene and other aromatic products (just less benzene) were determined with an HP-5 capillary column and a flame ionization detector (FID).

3. Results and discussion

3.1 Characterizations of catalysts

From the XRD patterns of supports used in this study(Fig. 1-A),it was obvious that both Mn/MgO-IM and Mn/MgO-PR supports had distinct diffraction peaks of Mg_6MnO_8 which derived from the rock salt structure of MgO by replacing one-eighth of the Mg^{2+} ions with Mn^{4+} ions and one-eighth with vacancies,[8]where Mn^{4+} ions appeared at the support surface and reacted with MgO to form Mg_6MnO_8,[9] also, due to this formation, MnO could disperse by one single layer on MgO surface.[10] In XRD spectra, there were some diffraction peaks (ca. 42.9°and 62.3°) overlapped together which indicated that both Mg_6MnO_8 and MgO were present on Mn/MgO-IM and Mn/MgO-PR supports. The size of Mg_6MnO_8 particles in either Mn/MgO-IM or Mn/MgO-PI was similar at 14-16 nm (Mg_6MnO_8 (311), $2\theta = 35.4°$). However, for MnO/MgO-MM support, besides the characterized peaks of MgO, the dominant phase of Mn species was α-Mn_2O_3 with a particle size of 32 nm (Mn_2O_3 (211), $2\theta = 32.9°$) which came from calcination of MnO_2.[11]

In XRD patterns (Fig. 1-B), a physical mixture composed of 50wt.%NiO and 50wt.%MgO without calcination exhibited evident difference of NiO and MgO signals in the range of 70-80°, so this region can be used to determine NiO and MgO.[12] However, in Ni/MgO catalyst, no NiO phase was detected, and the diffraction peaks were close to pure MgO, which indicated NiO dispersed well on the support of MgO owning to solid solution structure between NiO and MgO.[12] In Ni/MnO catalyst prepared by co-precipitation, there were also no NiO peaks besides α-Mn_2O_3 with a particle size of 58nm (Mn_2O_3 (211), $2\theta = 33.1°$), which suggested that Ni species also had a good dispersion.

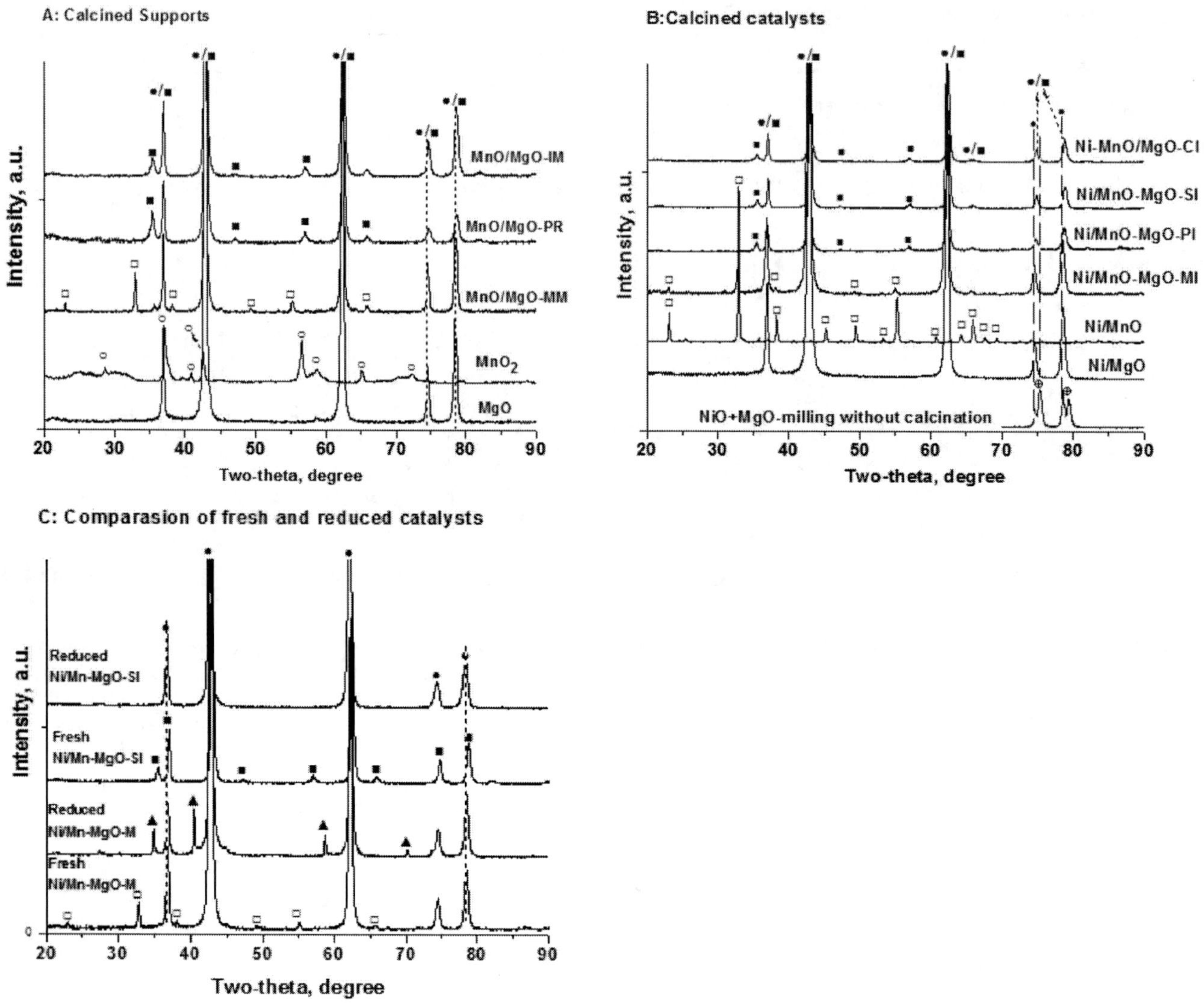

Fig. 1 XRD patterns of (A) Calcined supports, (B) Calcined catalysts, (C) Reduced catalysts. Symbols: □Mn_2O_3, ○MnO_2, ▲MnO,⊕NiO, ●MgO/or Ni-Mg-O, ■Mg_6MnO_8.

For the catalysts made by $Mn(NO_3)_2$ as the start precursor (SI, CI, PI), the crystalline phases of Mn and Mg species in XRD patterns did not change obviously in which $Mg_6MnO_8$and MgO were the main materials. But because the catalysts were calcined at 700℃ for 6h again, the particle size of Mg_6MnO_8 was increased slightly from 14-15nm to 18-20nm. Of particular interest in fig.1-B was that there were no NiO peaks present on catalysts, which might because NiO was covered byMg_6MnO_8 phase. As for Ni/MnO-MgO-MI catalyst, α-$Mn_2O_3$was also the main Mn species with a particle size of 36nm, no NiO and other manganese-contained phases.

In sum, Mg_6MnO_8 presented on all the catalysts with $Mn(NO_3)_2$ as the start precursor, whereas, Mn_2O_3 was the main phase of manganese on Ni/MnO-MgO-MI catalyst. So Ni/MnO-MgO-SI can be selected as a representativecatalyst to compare with NiO/Mn-MgO-MI.

When these two fresh samples were reduced at 700°C for 1h in H_2, the main states of Mn changed (Fig. 1-C). It was clear that Mg_6MnO_8 disappeared in the reduced Ni/MnO-MgO-SI catalyst without other Mn species, which suggested that the reduced Mn species would have a good dispersion on MgO surface. However, Mn_2O_3 was reduced to MnO on Ni/MnO-MgO-MI after reduction.

TEM pictures gave very different morphology of the refuced catalysts in Fig. 2. Fig. 2-A was the reduced Ni/MnO-MgO-SI, it was clear there were some squares on the MgO surface which was not showed in Fig. 2-B of the reduced Ni/MnO-MgO-MI. The different images derived from the different Mn species definitely. Noted that, in the EDS analysis of reduced Ni/MnO-MgO-MI, the Mn element seemed absent, that was because MnO has a high density which easily deposited in ethanol before TEM characterization.

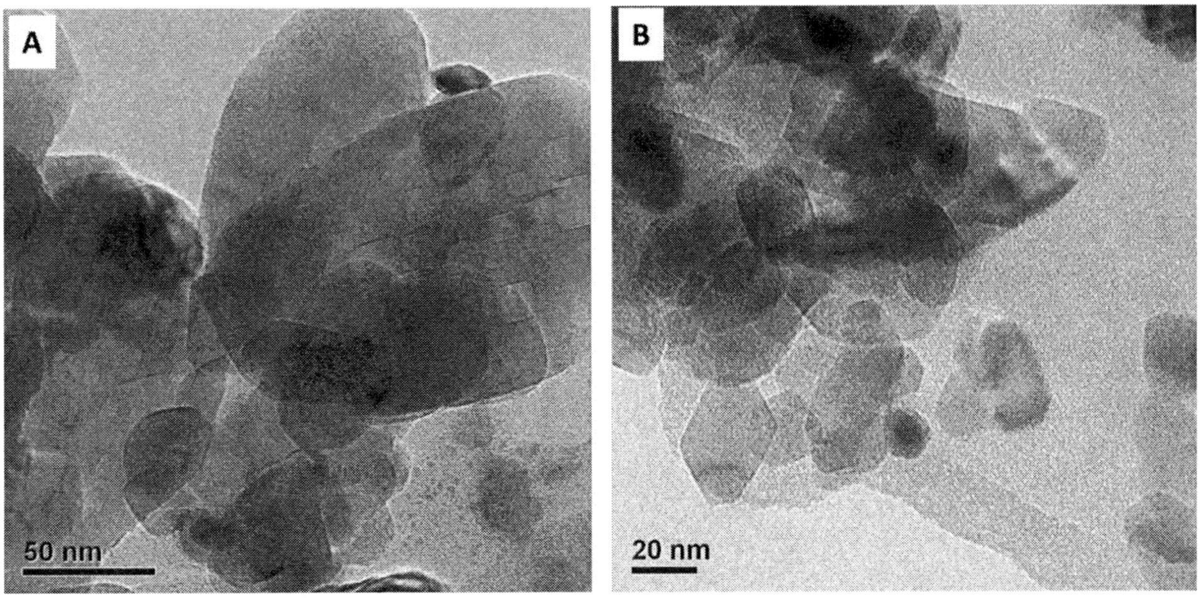

Fig. 2 TEM pictures of(A) reduced Ni/MnO-MgO-SI (B) reduced Ni/MnO-MgO-MI.

3.2 Activity of the employed catalysts in CO_2 reforming of toluene

The performances of catalysts mentioned above were drawn in Fig. 3. As (is) seen from these curves, over Ni/MgO catalyst without any promotes, the initial conversion of toluene can reach to 72.5% (data at 30min), comparably, it will be just 37% when just MnO severed as the support. However, for the Ni-MnO-MgO catalysts with $Mn(NO_3)_2$ as the precursor, the conversions of toluene were no more than 20%, among them, Ni/MnO-MgO-PI had a relatively higher activity with 22.5% of toluene converted at initial time, but it decreased quickly to less than 10% after 5h test. Interestingly, by using MnO_2 as the precursor, a highest and stable performance with 81% of toluene converted can be obtained over Ni/Mn-MgO-MI catalyst. For Ni-MnO-MgO with $Mn(NO_3)_2$ as the precursor, the total conversion of toluene were no more than 10%, including （that） 1.5%-2.5% of toluene was converted into benzene, they had a poor capacity for deep cracking of benzene. However, the proportion for toluene to benzene was also 0.8 % over Ni/MgO, 0.4% over Ni/Mn-MgO-MI, it can be considered that the existence of Mn promotes the conversion of toluene.

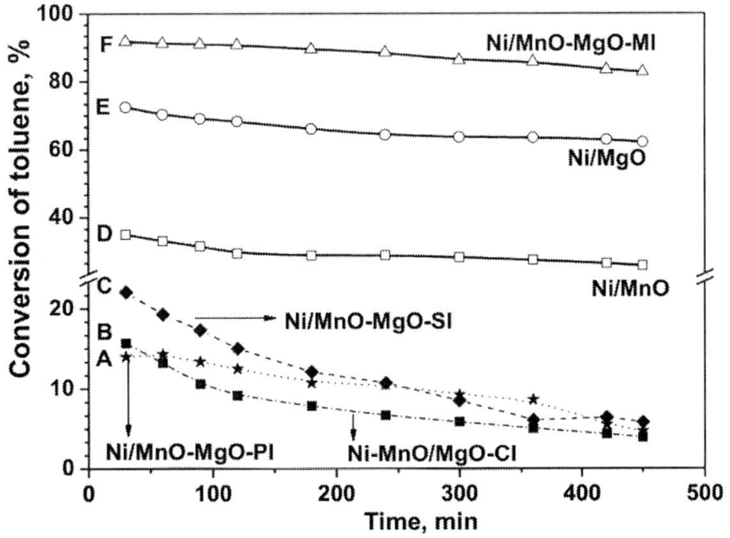

Fig. 3. The conversion of toluene over the catalysts on stream. A, Ni-Mn/MgO-PI; B, Ni/Mn-MgO-CI; C, Ni/Mn-MgO-SI; D, Ni/MnO; E, Ni/MgO; F, Ni/Mn-MgO-MI. Reforming conditions: 0.5ml catalyst, 0.005ml/min toluene, 14.5ml/min CO_2, 285ml/min Ar, GHSV= 36000 h^{-1}, temperature was 570℃.

Conclusion

Mn-modified Ni/MgO was investigated for CO_2 reforming of toluene. It was found that the preparation method played a crucial role in catalytic performance. Using $Mn(NO_3)_2$ as the precursor, Mg_6MnO_8 formed during calcination, which made the catalysts obtain a worse activity, however, using MnO_2 as precursor, the interaction between Mn species and MgO became weak, MnO can work as a support to distribute Ni particles, which benefited the reforming reaction.

Acknowledgements

This research work was supported by the Hi-Tech Research and Development Program of China (2007 AA 05Z415) and the Natural Science Foundation of Zhejiang Province (Y407113).

Reference

[1] Coll, R., Salvado, J., Farriol, X., Montane, D., Fuel Process. Technol., 2001, 74, 19-31.

[2] Zhang, R.Q., Brown, R.C., Suby, A., Cummer, K., Energy Conversion and Management. 2004, 45, 995-1014.

[3] Garcı´a. L., Salvador, M.L., Arauzo, J., Bilbao, R., Energy Fuels., 1999, 13, 851-859

[4] Li, X.T., Grace, J.R., Lim, C.J., Watkinson, A.P., Chen, H.P., Kim, J.R., Biomass Bioenergy. 2004, 26,171-193

[5] Kong, M., Fei,J.H., Wang, S., Lu, W., Zheng, X.M., Bioresour. Technol. 2011, 102, 2004-2008.

[6] Nandini, A., Pant, K.K., Dhingra, S.C., Appl. Catal. A: Gen. 2005; 290: 166-174

[7] Seok, S.H., Choi, S.H., Park, E.D., Han, S.H., Lee, J.S., J. Catal. 2002, 209, 6–15.

[8] Taguchi, H., Nagao, M., J. Mater. Sci. Lett. 1991,10, 658 - 659

[9] Guilliat, I.F., Brett, N.H., J. Chem. Soc. Faraday Trans. 1972, 68, 429.

[10] Yao, J.S., Meng, M., Luo, J.Y., Zha, Y.Q., Xie, Y.N., Hu, T.D., Liu, T., J. Mol. Catal. (China), 2006, 20, 300-305

[11] Ardizzone, S., Bianchi, C.L., Tirelli. D., J. Electroanal. Chem. 1997, 425, 19-23.

[12] Ruckenstein, E., Hu, Y.H., Appl. Catal. A: Gen. 1999,183, 85-92.

(62d) Direct Inter-Processes Integration for Energy Efficiency Optimization and GHG Emissions Reduction in Oil Refining

Mahmoud Bahy Noureldin, Process and Control Systems, Saudi Aramco, Dhahran, 31311, Saudi Arabia and Faisal AlMousa, Process and Control Systems Department, Saudi Aramco, Dhahran, Saudi Arabia

Industrial communities are invited to conserve energy resources not only to give a chance to pre-developing countries and future generations to grow and prosper but also to preserve the environment. Energy efficiency optimization is a quick answer to energy-based GHG emissions reduction. It enables us get two targets with one arrow. From GHG emissions outlook, energy efficiency optimization applications in industry can help us get out tens of thousands of cars abstractly from our roads and practically from our air every year.

Since the emanation of the pinch technology and its evolution to pinch analysis technique and the adaptation of MINLP techniques for process synthesis of mega complexes, the direct inter-processes integration has been considered impractical. For inter-processes "energy" integration many arguments against the direct inter-processes integration are certainly valid.

Plants are normally have different start up and shut down times; the processes can work at partial loads; the processes can have seasonal changes in its operating conditions; utility systems, heaters, coolers and HEN capital will not be reduced and may be even increased due to differences in processes/plants schedule and operation philosophy; the disturbance in one process plant can propagate to another one if they are mass and/or energy integrated; the distance-time/velocity lags in process will affect the controllability of integrated processes; the geographical distances among processes will cost us energy in pumping or compression and capital in piping; plants' safety might be impacted due to the travel of a fluid from one "hazardous" area classification to another; the fear of leakage and so on. Therefore, direct inter-processes integration while is beneficial to energy conservation and the GHG emissions reduction as will be demonstrated in the paper, is still to date almost ignored in the industrial community. Due to those concerns most of the current methods for inter-process integration are indirect and conducted using buffer systems. Buffer systems are either steam system, boiler feed water, hot water system, hot oil system or sometimes combinations of them. The industrial community does agree that, direct integration approach in inter-process integration (between several plants) is efficient and can render more saving in energy consumption and energy-based greenhouse gas emissions beyond the maximum possible using intra-integration only. In this paper, we demonstrated that huge potential for energy consumption and GHG emissions reduction in oil refining can be attained through smart integration among processes. The paper has two folds of results. First, while we agree with the validity of some of the arguments against the direct inter-processes/plants integration we believe and will demonstrate that we still have room for improvement in energy efficiency enhancement and energy-based GHG emissions reduction. We will identify at the targeting phase best possible scenarios for inter-processes integration and the most cost effective solution(s) now and in the future via "Plants' Smart Matching", without the discouraging solutions which require many inter-connections among plants as sometime reported in literature. Second, the intuition that mandates sophisticated inter-plants integration to achieve good benefits, the more you integrate among the processes (new mega sites consists of integrated refining, chemical, petrochemical and plastics plants), the better you save in energy consumption is not always true. The paper presents those two results and demonstrates the benefit that could be obtained from the direct inter-processes integration through two industrial case studies.

(62e) Energy Efficient Control Strategy to Operate Post- Combustion CO2 Capture in Response to Power Plant Load Reduction

Sepideh Ziaii, Gary T. Rochelle and Thomas F. Edgar, Department of Chemical Engineering, The University of Texas at Austin, Austin, TX

Energy Efficient Control Strategy to Operate Post- Combustion CO_2Capture in Response to Power Plant Load Reduction

Increasing concern about global warming has motivated the development CO_2capture and sequestration as a potential option for reducing emissions from PC and IGCC plants. One of the mature technologies that can capture CO_2emissions from a coal-fired power plant is amine-based absorption/stripping, which has the advantage of being added to an existing power plant. However, this technology is a very energy intensive, which can reduce the efficiency of power plants by 30%.

In thistechnology the aqueous amine solutioncountercurrently$_2$from the flue gas by exothermic chemical reactions and significantly reduces the CO_2concentration in the gas stream exiting the absorber. The CO_2-rich amine solution from bottom of the absorber is sent through a cross exchanger and directed to the stripper. The stripper temperatureis maintained$_2$reaction and liberate CO_2. The lean solution leaving the bottom of the stripper iscooled and $_2$. The concentrated CO_2stream from the top of stripperis compressed

Due to high-energy demand that is mostly attributed to solvent regeneration and CO_2compression, many research efforts seek to find the most energy efficient steady sate design . However there are few studies havebeen performed on minimizing energy consumption during possible operational scenarios for an existing capture plant. This study has simulated

This work incorporated models of MEA absorption/stripping process, CO2flowsheet in the framework of Aspen Custom Modeler. Absorber and stripper are packed columns modeled as multi-segment components with time variant mass and energy balance equations. Each segmentis modeledwas modeledare modeled

During boiler load variation in power plants, so many process variables are subject to change. One of the important variables influencing downstream capture plant operation is the property of flue gas entering the absorber. Based on discussion in literature on reduced load operation of coal-fired power plants, itcould be assumed that as load reduces, the rate of flue gas is reduced accordingly while the composition and temperature are constant. The other issue resulting from power plant load change is related to thesteam rate1st

Regarding energy efficient operation, the simulation results demonstrated that a variable speed compressor is much more beneficial than a constant speed one. Adjusting compressor speed reduces lost work by pushing the compressor away from surge condition and avoids additional energy use due to recycling gas in the compressor. In addition, as total steam rateis reduced in power cycle, the reboiler steam pressure and flow rate is reduced and consequently CO22heat duty equivalent work; however, the temperature should not exceed 1202is limited by maximum compressor speed, which is typically 120% of rated speed.

Another potential processvariable which can be manipulated to minimize reboiler lost work and save the compressor from surging is solvent circulation rate. At fixed reboiler temperature, there is an optimum solvent rate that minimizes lost work and maximizes CO2is set by compressor and pump operability. The minimum flowis always setis operated

Based on minimization of total lost work by compressor speed and solvent circulation rate performed in the presence of operational constraints associated with pumps, compressor and solvent thermal degradation, this study gives a practical optimum CO_2removal over 40%-100% operating load range. For this range of load change in an MEA plant designed for 90% removal, the plant can remove up to 94% of inlet CO_2. The practical minimum removal over that load range is about 85%, where the reboiler temperature reaches its maximum allowable value and the lost work is at a minimum.

Reducing the load of power plant that reduces extracted steam rate in the capture results pushes the compressor operating curve to the surge limit such that at minimum load (40%) , adjusting solvent rate and compressor speed no longer moves the operating curve away from

Based on energy minimization in response to power plant load reduction, two control strategies were found to operate the capture close to minimum energy path at each desiredremoval .

1. Ratio control between reboiler steam rate and CO_2rate in rich solution or
2. Controlling lean loading to vary proportional to the CO_2removal change.

In the future we plan to observe and compare dynamic performance of these two control strategies.

(62f) Study on Cascade Refrigerant and Mixed-Refrigerant Systems for Ethylene Plants

Ha Dinh, Jian Zhang and Qiang Xu, Dan F. Smith Department of Chemical Engineering, Lamar University, Beaumont, TX

Refrigeration system holds an important role in most chemical/petrochemical processes. A refrigerant system transfers heat from low-temperature sources to high-temperature sinks through vapor-compression cycles at the expense of mechanical work, magnetism, laser or other means. In many industrial processes, the refrigeration systems are employed for chilling and freezing the final liquid/solid products or liquefying the gas intermediates/products for the downstream cryogenic separation. Optimal synthesis of these systems is attractive not only because of their economic benefits but also because they incorporate most of energy transfer operations in widespread use in the chemical industry. The traditional cascade refrigeration system (CRS) used in ethylene plants includes multiple refrigerants and multiple temperature and pressure levels. Each component subsystem includes a compression section, a condensing section and several temperature levels with sub-coolers, expansion valves, evaporators and flash drums. Recently, mixed-refrigerant system (MRS) has shown some advantages comparing to the single-component one. A MRS uses a mixture as the refrigerant instead of several pure components as in conventional cascade refrigeration systems. This system operates within smaller temperature differences at the lower limit; therefore, it leads to a smaller increase in entropy and consequently a smaller loss in energy. Besides, an MRS also requires a simpler unit configuration and fewer maintenance problems.

In this study, both types of refrigeration systems used in ethylene plants are studied based on exergy analysis, a powerful method to explore system energy loss below ambient temperature. The procedure involves three consecutive steps: Thermodynamic Analysis, Synthesis Model Development and Solving, and Solution Validation and Evaluation. Rigorous steady-state models of both refrigeration systems are firstly constructed and employed to generate Exergy-Temperature (B-T) chart in order to analyze their thermodynamics and energy distributions. An exergy-based mixed-integer linear programming (MINLP) model is developed to maximize the energy efficiency of each system, where multiple refrigerants with multiple recycling loops are simultaneously addressed and all of cooling demands are fulfilled. The optimal solution is subsequently examined by conducting rigorous simulation to check their feasibility and consistency. Economic evaluation is also carried out from the calculated capital cost of new equipments and annual operating cost. The efficacy of the developed methodology is demonstrated by two case studies: a cascade refrigeration system and a mixed-refrigerant system used in two ethylene plants.

Operations Optimization of Concentrating Solar Power Plants I: Challenges and Opportunities

I. Salbidegoitia and I. Mendicute
Torresol Energy
Madrid, Spain

G. Deiu, A. Trenchard, J. Visschers and R. Nath
Honeywell Process Solutions
Europe and Americas

Torresol Energy
reinventing solar power

Honeywell

AIChE Spring Meeting
April 3, 2012

Agenda

- Introduction
- Operations Optimization
  - Objective
  - Opportunities
  - Challenges
- Conclusions

AIChE Spring Meeting
April 3, 2012

Introduction: Torresol

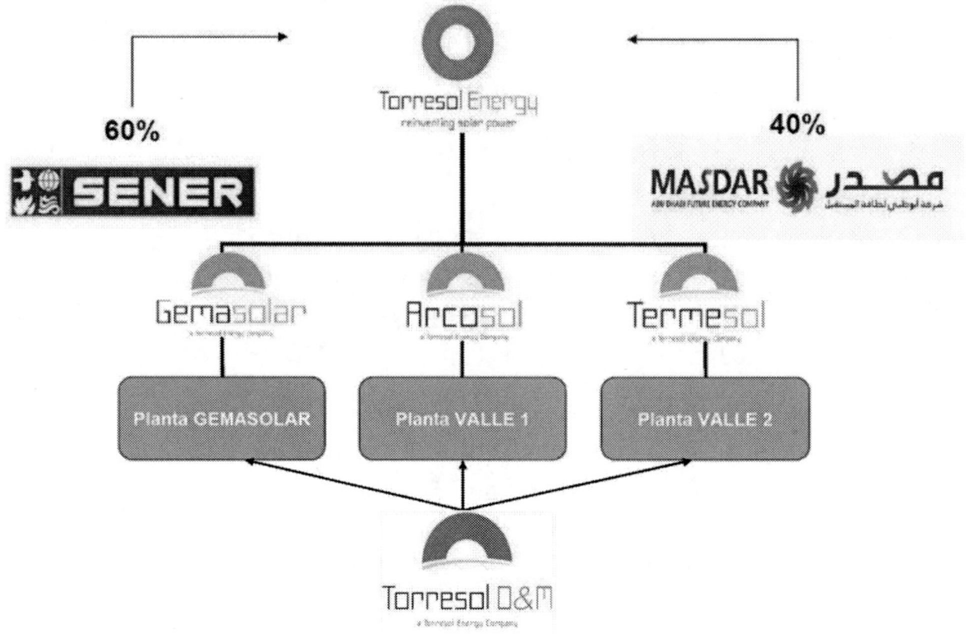

Introduction: Solar Energy

- Produced by thermonuclear reaction in the sun
- Produces entire spectrum of EMR

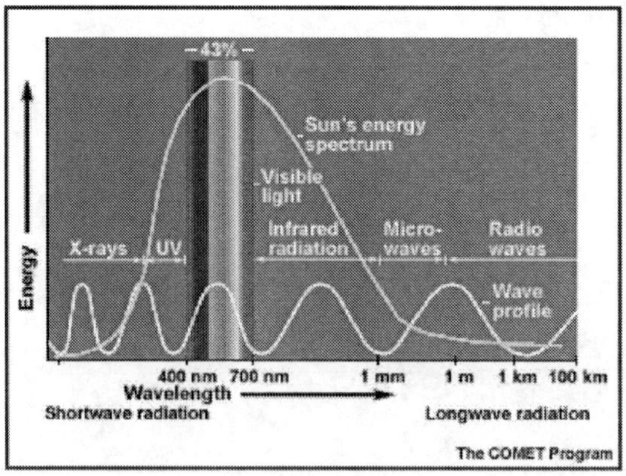

Introduction: Solar Energy

- PROs
  - Abundant
  - Clean

- CONs
  - Diffused
  - Cyclical

- Solution
  - Concentration
  - Storage

Introduction: Solar Energy Concentration

- Central or Power Tower

- Parabolic Troughs

Gemasolar

Valle 1 & Valle 2

Introduction: Solar Energy Concentration

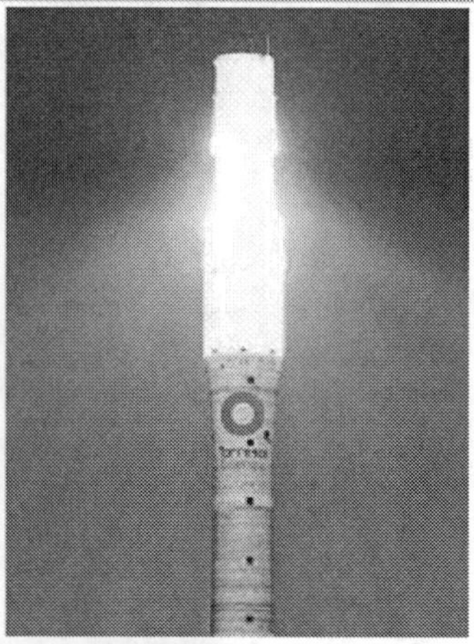

Introduction: Solar Energy Storage

Valle 1 Salt Storage System

Introduction: process diagram

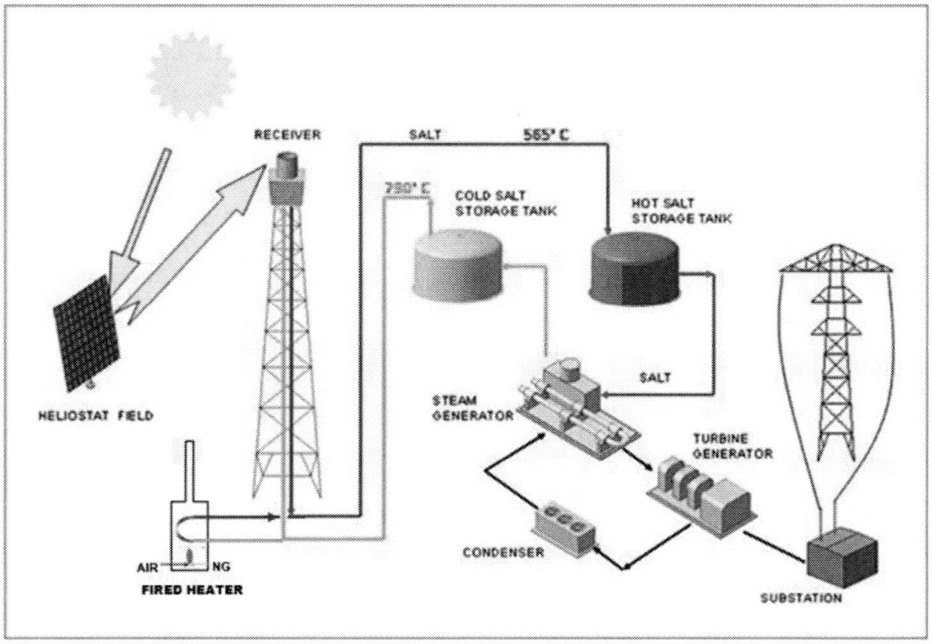

Introduction: economics

High Capital cost, mitigated by

- Federal and state tax incentives
- Green energy pricing incentives
- Use of Fired Heater (limited to an annual maximum)

Operations Optimization – objective

Maximize profit over the optimization time horizon.

$$\approx \text{MAX} \sum_{t=0}^{t=EOY} c\_eee_t * EEE_t + c\_eeeod_t * EEEOD_t - c\_eei_t * EEI_t - c\_fei_t * FEI_t - p\_cos * COS_t$$

where

c_eee_t	is the credit for electrical energy export
c_eeeod_t	is the credit for on-demand electrical energy export
c_eei_t	is the cost of electrical energy import
c_fei_t	is the cost of fuel energy import
p_cos	is the cost per transition
EEE_t	is the amount of electrical energy export
$EEEOD_t$	is the amount of on-demand electrical energy export commitment
EEI_t	is the amount of electrical energy import
FEI_t	is the amount of fuel energy import
COS_t	is the indicator for transition, 0/1

AIChE Spring Meeting
April 3, 2012

Operations Optimization – objective

Hour to hour variation

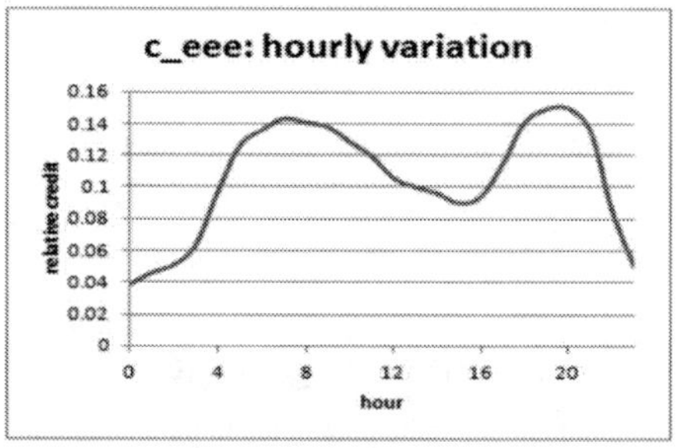

Month to month variation

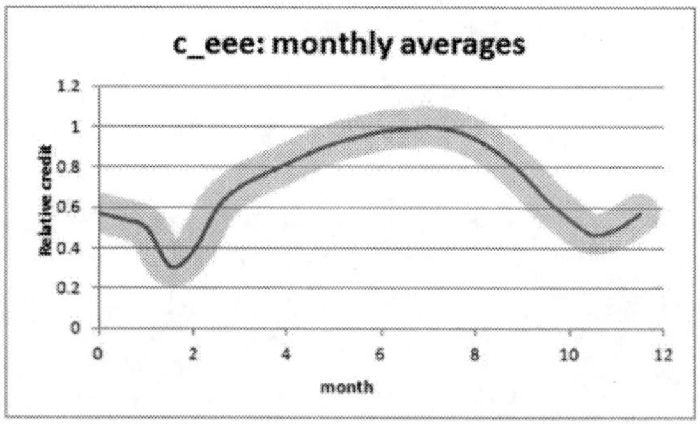

AIChE Spring Meeting
April 3, 2012

Operations Optimization – opportunities

Opportunities:

- Flexibility in production
- Flexibility in use of Fired Heater

Operations Optimization – opportunities (1)

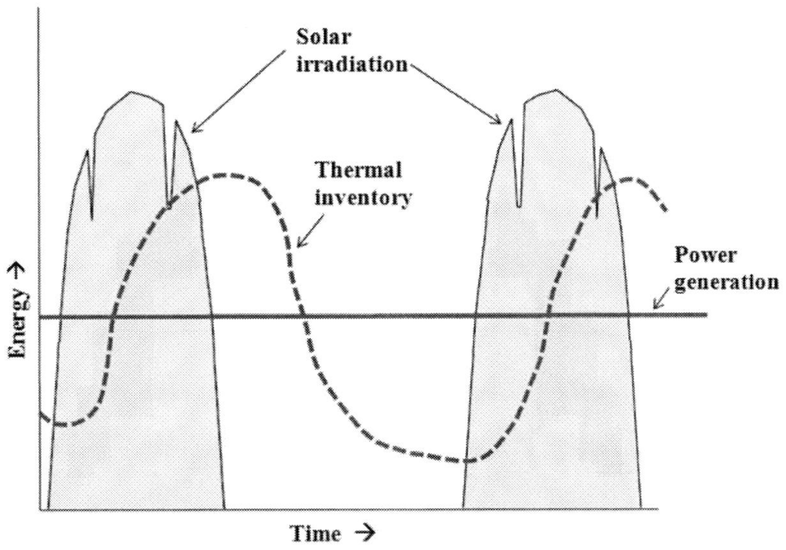

Operations Optimization – opportunities (2)

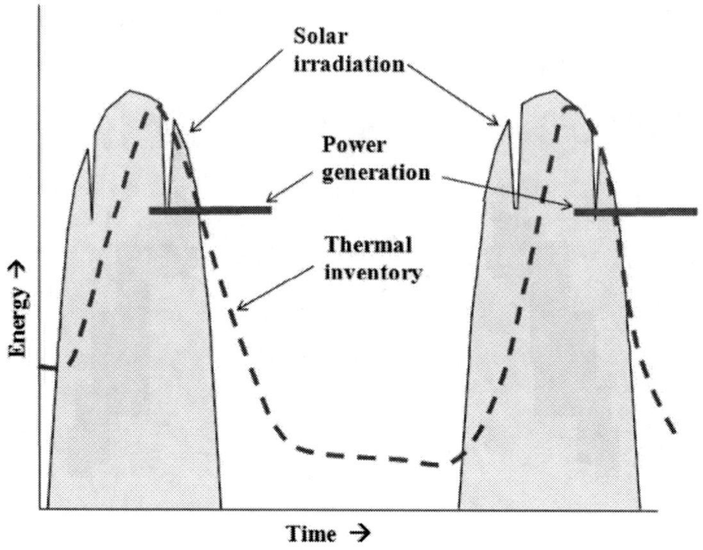

AIChE Spring Meeting
April 3, 2012

Operations Optimization – opportunities (3)

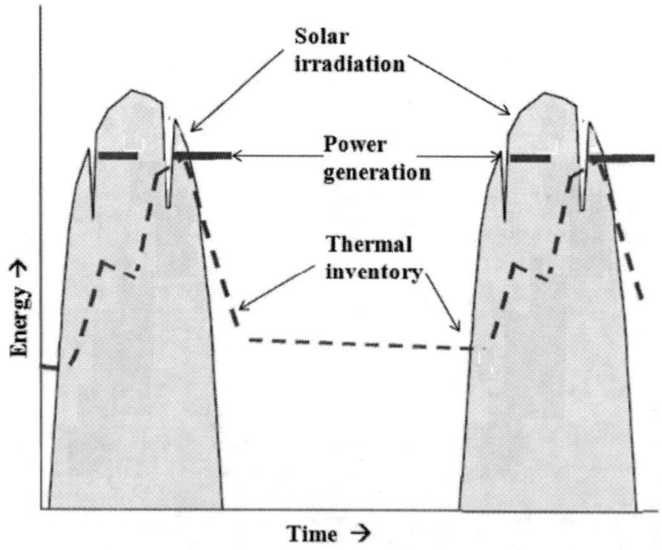

AIChE Spring Meeting
April 3, 2012

Operations Optimization – opportunities (4)

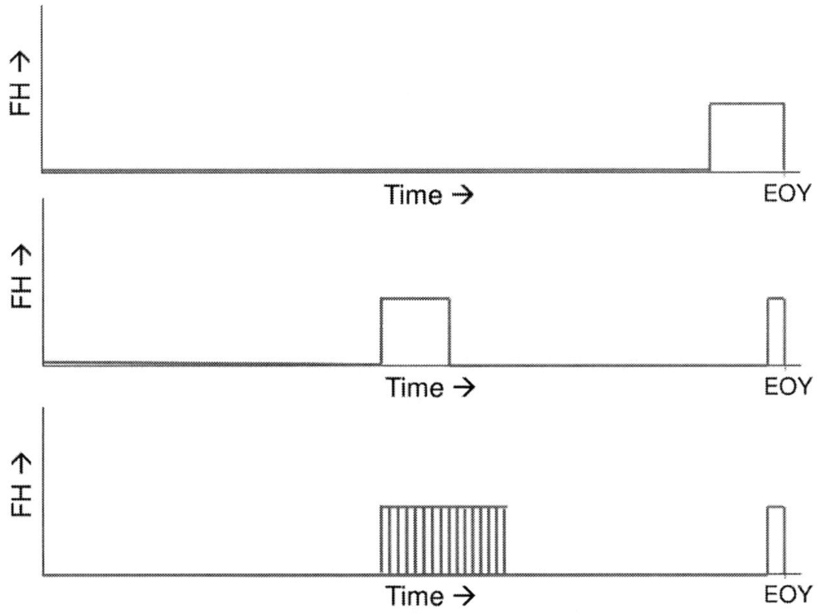

AIChE Spring Meeting
April 3, 2012

Operations Optimization – modeling challenges

Modeling challenges:

- Number of time periods
- Number of time periods with integers
- Complex electrical market
- Change of state penalties
- Cost of imported power
- Annual Fired Heater constraint

AIChE Spring Meeting
April 3, 2012

Operations Optimization – modeling challenges

Number of time periods

- Hourly time slices: 8760 periods
- Could do time lumping for more distant periods

Operations Optimization – modeling challenges

Number of time periods with integers

- Integers are required to model
  - Operating ranges
  - start up/shut down events
- Integers are computationally demanding
- Be judicious in use of integers: limit it to short term that includes the nomination time horizon

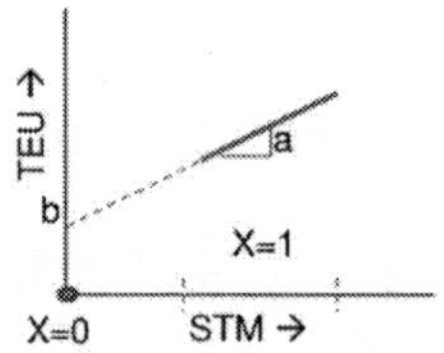

TEU = a * STM + b * X

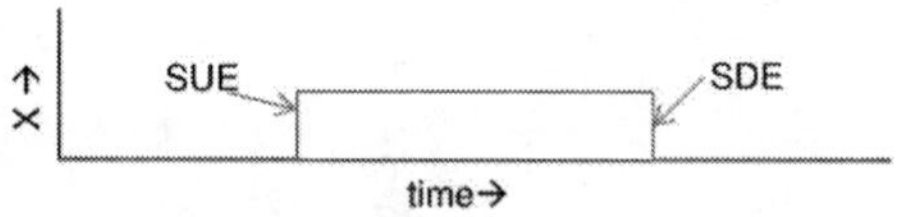

Operations Optimization – modeling challenges

Complex electrical market

- Daily, multi-round auction process
- Time dependent price
- Deviation penalties after initial nomination (smaller)
- Deviation penalties after final nomination (larger)

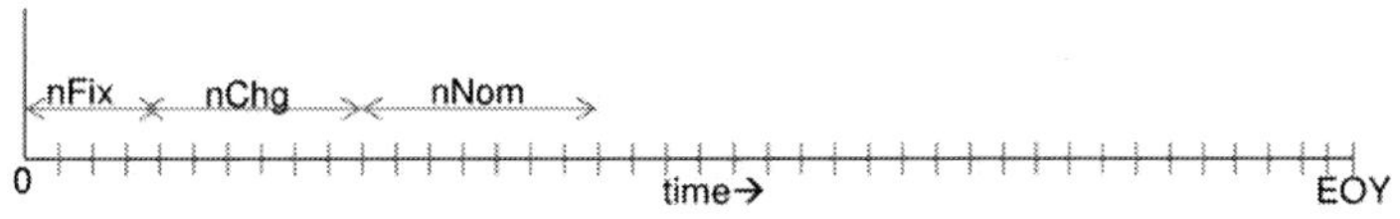

$$= \text{MAX} \sum^{EOY} c\_eee_t * EEE_t - \sum^{nFix} p\_dev_L * \Delta EEE_t - \sum^{nChg} p\_dev_S * \Delta EEE_t \ldots$$

$$\Delta EEE_t = |\ EEE_t - nom_t\ |$$

Operations Optimization - modeling challenges

Change of state penalties

- For CT, FH and PB operations
- Penalize start up and shut down events

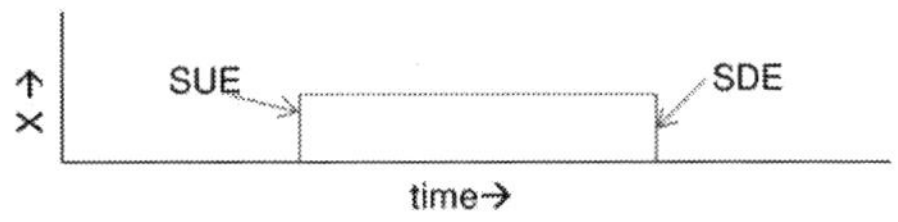

$$= \text{MAX} \ldots - \sum^{EOY} p\_sue * (CT\_SUE_t + FH\_SUE_t + PB\_SUE_t\ )$$
$$- p\_sde * (CT\_SDE_t + FH\_SDE_t + PB\_SDE_t\ ) \ldots$$

- Add transition energy use and delays, if necessary

Operations Optimization - modeling challenges

Cost of imported power

- Energy cost
- Demand cost
- Excess demand penalty

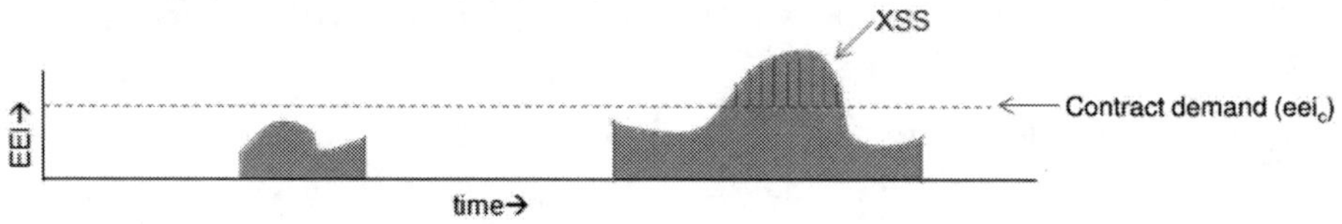

$$= \text{MAX} \ldots - \sum^{EOY} (c\_eei_t * EEI_t + c\_eed * eei_c + p\_eed * XSS_t \ldots$$

Operations Optimization - modeling challenges

Fired Heater usage constraint

- Limits total TEG as a % of total EEG (fh_{max})

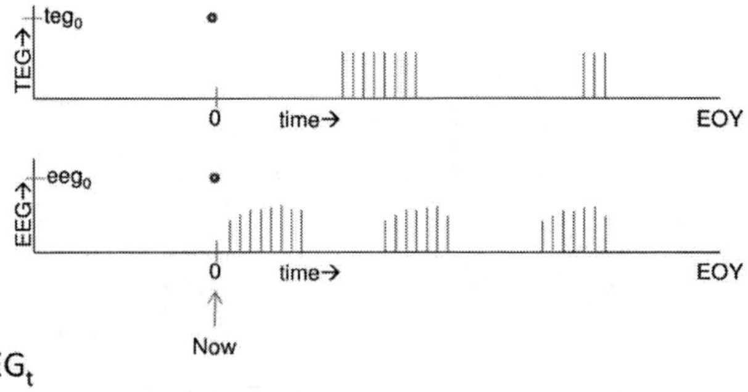

$$TEG = teg_0 + \sum_0^{EOY} TEG_t$$

$$EEG = eeg_0 + \sum_0^{EOY} EEG_t$$

$$TEG < fh_{max} / 100 * EEG$$

Operations Optimization - challenges : summary

Realistic models can be developed by addressing each of the modeling challenges.

Expect the model to be large;

- 10^3 time periods
- 10^5 continuous variables
- 10^5 constraints
- 10^2 integer variables

Target execution time < 10 minutes

Conclusions

Operation of concentrating solar power plants offers unique flexibilities and opportunities for significant profit improvement.

Large scale, multi-period, scheduling optimizer would be an essential tool to fully exploit these flexibilities.

Work is in progress towards design and implementation of operations optimizers for the three Torresol CSPs ... expected completion within the next 12 months.

Q & A

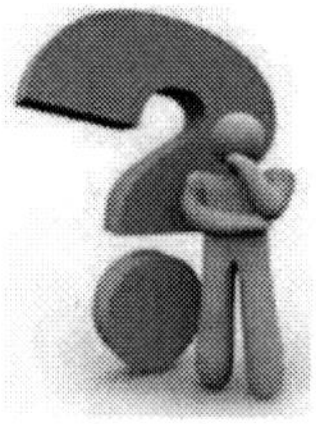

AIChE Spring Meeting
April 3, 2012

(91b) Numerical Simulations and Experiments on a Photovoltaic Solar Wall

Himanshu Dehra, 1-140 Avenue Windsor, Lachine, QC, Canada

1. Numerical SimulationsA two dimensional thermal network model is proposed to predict the temperature distribution in the photovoltaic solar wall section of an outdoor test room facility at Concordia University, Montr"al, Qu"bec, Canada. Steady-state thermal network nodal equations are developed for a photovoltaic solar wall section by neglecting thermal storage in double-glass photovoltaic module and well-insulated back panel. The enclosure analysis is assumed for obtaining radiative conductances to ascertain the effect of radiative heat exchange in between the distant composite surface nodes. One dimensional solution obtained from the heat balance model eliminates the non-linear variation of temperature dependent boundary conditions applied to thermal network. The proposed model was able to predict the temperature distribution in the test section for a range of measured climatic data under the strong influence of conjugate heat exchange.

2. Experiments The preliminary field experiments were conducted on ascertaining electrical and thermal characteristics of a pair of glass coated photovoltaic (PV) modules. Both indoor and outdoor experiments were conducted on glass coated PV modules. A pre-fabricated outdoor room was also used for conducting some of the outdoor experiments on glass coated PV modules installed on a wooden frame. The experiments on obtaining electrical characteristics of PV modules included were indoor solar simulator tests conducted under standard operating conditions, outdoor tests conducted with and without a wooden frame under available climatic conditions. Both indoor and outdoor experiments were conducted for obtaining current-voltage and current-power characteristics for a pair of PV modules. The current-voltage measurements were also obtained for determining electric power output with a series electrical circuit connection of a pair of vertically inclined PV modules installed on a wooden frame. The experiments on determining thermal characteristics included were outdoor tests conducted for obtaining temperatures, air mass velocities, sensible heat, thermal storage and thermal time constants of PV modules with and without wooden frame. After conducting a series of field experiments, preliminary results have concluded that: i) glass coated PV module installed on a wooden frame has more heat generating capacity in comparison to a window glass or a stand alone PV module; ii) the generation of electric power is affected with operation of vertically inclined PV modules; and iii) electrical and thermal characteristics are function of both ambient air temperature and thermal storage losses.

(91c) DFT Study on Hydrogen Desorption From NaMgH3 (001) Surface

Fernando A. Soto, Louisiana Tech University, Ruston, LA and Daniela S. Mainardi, Chemical engineering, Louisiana Tech University, Ruston, LA

DFT Study on Hydrogen Desorption from $NaMgH_3$(001) Surface

Authors: Fernando Soto[1], Daniela S. Mainardi[2]

[1] Institute for Micro-manufacturing, Louisiana Tech University

[2] Institute for Micro-manufacturing, Chemical Engineering, Louisiana Tech University

$NaMgH_3$ is the only magnesium based ternary hydride with Na and it has attracted the attention as a possible candidate for hydrogen storage material because it has high gravimetric hydrogen density (6 %) and high volumetric hydrogen density (88 kg/m_3). Hydrogen disassociation in this material occurs at atmospheric pressure and almost 670 K, which is well above the operable range for on board hydrogen storage applications and a major impediment for practical applications. In this study, aiming to enhance the dehydrogenation process on transition metal modified- $NaMgH_3$,(001) surface has been investigated by using density functional theory calculations. For the surface calculations, three types of models can be identified: Hydrogen desorption from the surface of the $NaMgH_3$ (H_{surf}), within the $NaMgH_3$(H_{bulk}), and at the bulk/surface interface ($H_{bulk/surf}$).

We propose that Hydrogen desorption process is facilitated in$NaMgH_3$by doping the material with Ti at the surface site and Zn as a co-dopant at the bulk site. The dehydrogenation energy of the co-doped surface structure is low, in comparison to the pristine model, Furthermore, when doped with transition metals from the 3d and 4d block elements, the dissociation barrier can also be lowered, in some cases with better results than Ti and Zn. Carbon doping was found to be the most promising, based on the considerations of the activation barrier, and dehydrogenation energy cost. Our results predict that these typesof model structures are potential useful materials for hydrogen storage application.

(91d) The Reaction Pathways of Glycerin in Supercritical Water

Jared S. Bouquet[1], Ryan E. Tschannen[2], Aaron C. Gonzales[2]and Sunggyu Lee[2], (1)Department of Chemical & Biological Engineering, Missouri University of Science and Technology, Rolla, MO, (2)Chemical and Biomolecular Engineering, Ohio University

Glycerin, a byproduct of biodiesel production, is a cheap and widely available feedstock that may be used for hydrogen production in energy application. One method that can be used to produce hydrogen from glycerin is using supercritical water as a reaction medium. The process is non-catalytic and can handle a wide range of impurities. The primary gaseous products of glycerin in supercritical water are hydrogen, carbon monoxide, methane, carbon dioxide, and ethane. In order to optimize this process, a thorough understanding of the reaction pathways of glycerin in supercritical water is needed. The different mechanistic pathways of glycerin in supercritical water were examined, including decomposition, recombination, pyrolysis, and reformation. Reactions occurring between product gases, such as the water gas shift reaction and methanation, were also examined. An experimental study was also performed in order to evaluate the accuracy of the different reaction pathways of glycerin using a 0.1L Haynes® Alloy 282 reactor system, and particular attention was given to the role of supercritical water in the reaction chemistry. Experiments were done at a variety of temperatures, space times, and water-to-glycerin molar ratios to gain a comprehensive understanding of the reaction pathways of glycerin.

(91e) Glycerol to Oxygenates and Chemicals in Supercritical Fluids

Maoqi Feng, Chemical Engineering Dept., Div. of Chemistry and Chemical Engineering, Southwest Research Institute, San Antonio, TX and Chee Kai Tan, Div. of Chemistry and Chemical Engineering, Southwest Research Institute, San Antonio, TX

Glycerol (glycerin) is a byproduct from biodiesel production via transesterification reaction from renewable vegetable oils and animal fats. With the increase of biodiesel production in the world, the glycerol supply to the market is beyond the need for chemical derivatives. The economics of biodiesel production will be improved if value-added products such as fuels or oxygenated chemicals can be produced from glycerol. This paper will address supercritical process conversion of glycerol to higher value-added chemicals and fuel components under catalytic conditions.

(91f) Autothermal Reforming of Glycerol in a Dual Layer Monolith Catalyst

Yujia Liu, Dept of Chemical Engineering and Material Science, Stevens Institute of Technology, Hoboken, NJ and Adeniyi Lawal, Dept. of Chemical Engineering and Materials Science, New Jersey Center for MicroChemical Systems, Stevens Institute of Technology, Hoboken, NJ; Chemical Engineering and Materials Science Department, Stevens Institute of Technology, Hoboken, NJ

Autothermal Reforming of Glycerol in a Dual Layer Monolith Catalyst

Yujia Liu and Adeniyi Lawal

New Jersey Center for Microchemical Systems, Dept. of Chemical Engineering and Materials Science, Stevens Institute of Technology, 1 Castle Point on Hudson, Hoboken NJ 07030

Abstract

Glycerol is a by-product during biodiesel production via transesterification of triglycerides. The conversion of glycerol to synthesis gas and then to methanol represents an important niche for glycerol processing that could relieve small biodiesel producers from high glycerol disposal cost while reducing their dependence on methanol from fossil fuels. 2O/C) ratio, oxygen to carbon (O_2/C) ratio and gas hourly space velocity (GHSV). The optimum operating conditions to produce desired conversion of glycerol, yields of H_2, and H_2/CO ratio with minimal coke formation were also determined.

(120b) Simulation, Optimization and Costing of a Fully Integrated Multi-Product Biorefinery

Aryan Geraili Nejadfomeshi, Paritosh K. Sharma and Jose Romagnoli, Chemical Engineering, Louisiana State University, Baton Rouge, LA

Simulation, Optimization and Costing of a Fully Integrated Multi-Product Biorefinery

Aryan Geraili Nejadfomeshi, Paritosh Sharma, J.A. Romagnoli

Louisiana State University, Baton Rouge, LA, U.S.A

Over the last century, the energy consumption has increased progressively owing to the growing demand by the industrialized societies and the rising world population. The current global state of energy supply is highly dependent on fossil fuels. Yet, there are varieties of alternative methods to produce energy from renewable sources. In this project, we study the development, optimization and costing of a fully integrated multiproduct biomass refinery to convert agricultural residue to value-added chemicals and fuels.

Within the framework of this project, the present study includes the process simulation, the study of the costing equipment at different scales, and strategic design of supply chain and processing capacity under uncertain market conditions. In this presentation, the focus is on the simulation and optimization of a multiproduct biomass refinery by means of different technology choices and process superstructures.

For our case study, the prospective product portfolio consists of ethanol, butanol, lactic acid and succinic acid. A fermentation platform is used to convert biomass to products. The potential pretreatment technologies which are investigated include steam hydrolysis, dilute acid (DA), ammonia fiber explosion (AFEX). Fermentation configurations that are investigated include simultaneous hydrolysis with either (1) separate C5 and C6 fermentation, or (2) co-fermentation of C5 and C6 sugars. A multi-period deterministic process superstructure optimization model is used to arrive at the optimal product and technology configuration. AFEX and Co-fermentation are found to be the optimal technologies based on the process yield and capital and operating cost. The optimal product mixture includes ethanol and succinic acid. Once the optimization is performed for this process, the simulation of the optimal process configuration is carried out on different capacities.

Additionally, waste management technologies are investigated and cost is evaluated through simulation. These technologies are fed back into superstructure optimization model to determine the economic impact of investment in waste mitigation on the biorefinery. Three waste mitigation technologies are simulated; (1) Carbon capture and sequestration of fermentation gas (from ethanol fermentation) into succinic acid fermentation tanks (to improve succinic acid yields), (2) Carbon capture, purification, and transport from emitted flue gas (from power generation) for enhanced oil recovery markets, and (3) Anaerobic and Aerobic digesters to mitigate solid effluents from the production processes. The fermentation gas is essentially pure carbon dioxide which doesn't require additional purification and improves directly the bottom line of the biorefinery (reduction in raw material costs of CO_2). The flue gas is a mixture of gases and impurities which require further scrubbing and compression before commercial usage in enhanced oil recovery (although providing a potentially more lucrative market). Finally, the digesters provide biogas which can be used to improve power generation capabilities of the plant. After doing the superstructure optimization, it was found that anaerobic digestion and fermentation gas sequestration can be incorporated profitably with the production processes.

Keywords: Multi-Product Biorefinery, Process Optimization, Waste Mitigation, Carbon Capturing and Sequestration

(120a) A Life Cycle-Based Analysis and Optimization Framework for Designing Sustainable, Multi-Product Biomass-to-Bioproducts Value Chains

Paritosh K. Sharma, Chemical Engineering, Louisiana State University, Baton Rouge, LA and Jose A. Romagnoli, Chemical Engineering, Louisiana State Univeristy, Baton Rouge, LA

A life cycle-based analysis and optimization framework for designing sustainable, multi-product biomass-to-bioproducts value chains

In recent years there has been a marked surge in the search for alternative sources of energy that wean the world off of dependence on fossil fuels and reduce our carbon foot-print on this world. After a boom in U.S. corn-based ethanol in the early part of the 21st century, the interest has gradually shifted towards more viable sources of biofuels and biochemicals. Cellulosic ethanol, biodiesel, and syngas are examples of such fuels that are extremely attractive owing to the fact that the raw materials can be composed completely of "left-over" wastes of food crops and forest harvests that don't interfere with the human food chain and the natural ecosystem. Even with increased research emphasis on biofuel and biochemical production technologies, large scale production is still hampered by many factors. Some of these factors that have received considerable interest in recent times include, cultivating more robust feedstock, genetically engineering enzymes and microbes that are more efficient and resistant to poisoning by by-products and improving process technologies that can render large scale production of biofuels a more profitable option for private entities. One area that has not received enough research interest in recent times is value chain modeling and optimization in order to make the entire value chain more competitive in the energy sector. Modeling an enterprise that produces renewable-based fuels and chemicals can provide valuable insight into the inter-play of the bioproduct value chain. This idea had been motivated by a recent surge in the research areas concerning supply-chain optimization and how cost-cutting measures throughout a supply-chain can render an otherwise sluggish enterprise, profitable.

In the current formulation, we apply a distributed, systematic approach to model, optimize, and analyze three aspects of sustainability for a biomass-to-bioproducts value chain; business, environmental, and social sustainability. We assume a vertically integrated bio-enterprise that bears the tasks of contracting feedstock production and conversion, and final product distribution. Life cycle analyses are conducted to generate forecasts for environmental parameters, infrastructural and employment data, input costs and supplies, product yields, and output market conditions, which are integrated with integer-based stochastic framework that yield strategic decisions such as feedstock and supplier selection, transportation logistics design, technology and product selection, waste mitigation design, and capacity design under uncertainty. A two-stage optimization strategy is used to increase computational efficiency while extracting the biggest value drivers for the successful construction and operation of a biorefinery.

The first stage in optimization is termed*portfolio design*; a deterministic integer programming model is formulated to design the optimal superstructure of feedstock, conversion technologies, and products. Stakeholder value of the entire value chain, including suppliers, customers, capital providers, and the environment, is maximized in the model with a rolling-time horizon of 12 years. The model equations and constraints included are material, energy, capacity, cash, and debt balances, transformation equations, supply and demand constraints, and capacity and resource constraints. Carbon taxes and mitigation credits are used to represent environmental implications of biorefinery activities. The outputs of the model include integer decisions regarding feedstock, suppliers, technologies, and product selection. Furthermore, sensitivity analysis is carried out on the optimal value chain configuration to determine process, supply chain, and market parameters that are the greatest business and environmental value drivers for the biorefinery.

The second stage of optimization is formulated as a stochastic decision tree optimization model. The purpose of the model is to design a timing strategy for feedstock production and pricing, incremental technology acquisition, and capacity design given uncertainty in input costs and supplies, technology costs, and product demands and prices. The model is similar in form and structure to the deterministic MIP from the first stage with stochastic forecasts as inputs. The decision tree nodes are represented by binary integer variables and the tree is generated using a real options methodology. Two types of options are modeled; Sequential growth options to model when to expand capacity, and switching options to represent flexible biorefining systems with the ability to switch product volumes between different products. A binomial decision tree is used to discretize continuous stochastic forecasts for inputs. The final objective function is a probability weighted stakeholder value whose form is derived from the deterministic model. The screened portfolio choices that are obtained from the first-stage optimization are used as inputs. Integer decisions for the model include capacity establishment and expansion, investment in new technologies, and product switching by re-allocation of feedstock to different product groups.

The resultant framework is applied to a prospective biomass refinery in Southeastern United States. The feedstock choices include energy crops, namely, switchgrass, sorghum, and energy cane. The feedstock decisions are affected by contractual costs for land, fixed and operating costs for crop growth and harvest, transportation expenses, environmental impact, and biomass and sugar yields of the crop. The initial product portfolio includes ethanol and butanol as fuels, and succinic and lactic acid and value-added chemicals. The conversion and separation technologies are demarcated into three groups; pretreatment, hydrolysis (steam, dilute acid (DA), ammonia-fiber explosion (AFEX)), and fermentation ((1) separate C5 and C6 fermentation or (2) co-fermentation) which employ common unit operations for each product, and separation and product recovery (distillation, molecular sieves, pervaporation, membrane separation) which are specific to each product. The product-technology set choices are affected by capital costs, operating costs, process yields, environmental impact, and market potential.

The screening model results yield energy cane as the feedstock choice owing to high biomass yields and ease of processing, and ethanol and succinic acid as the products owing to great market potential and favorable profit margins. Dilute acid pretreatment followed by simultaneous hydrolysis and co-fermentation are selected as the conversion technologies owing to favorable capital cost structure and process economics including the energy and environmental load. Butanol and lactic acid are not chosen primarily due to the high cost of product recovery while distillation and molecular sieves (ethanol) and membrane separation (succinic acid) are chosen for the products that are selected.

The results are then fed to the second stage of optimization with uncertainty being represented in the marginal returns that are generated by capacity and production decisions. The uncertainty in the return implies uncertain revenue and cost streams. The individual components of the uncertain revenue and cost streams include uncertain product prices and uncertain input costs. The uncertain returns are assumed to capture all uncertainty during optimization while the distribution of returns is generated separately using the stochastic forecasts for the individual components. The results indicate an incremental capacity design to mitigate and manage capital and market risks with the overall expected stakeholder value at $MM 3.4 indicating that positive value is being created. The incremental capacity design is compared to economies of scale based capacity design where all capacity is established during the first planning year. The stakeholder value for economies of scale based design is $MM -30.3 indicating a $MM 33.7 advantage to using incremental capacity design. Product switching capabilities are recognized in the final two time periods (4 years) where a price shock to the upside in succinic acid prices is simulated. Capacity is re-allocated to succinic acid production to exploit the favorable market conditions with a 2% increase in the stakeholder value thus depicting the advantages of designing flexible production systems for biomass refineries.

(120c) Using System Dynamics to Support Consequential Life Cycle Analysis: The Forest Based "Drop-in" Biofuel Supply Chain Example

Nana Awuah Bortsie Aryee, School of Forest Resources(Research Group for Industrial Ecology, LCA and System Sustainability), University of Maine, Orono, ME and Anthony Halog, School of Forest Resources, University of Maine, Orono, ME

The biofuel research and production sector is enjoying an unprecedented amount of public and private effort in order to remove various financial and technical barriers. One of these very important barriers is the challenge of developing biofuels that are compatible with existing gasoline, jet fuel and diesel infrastructure. The United States Department of Energy has been championing efforts that ensure that the next generation of biofuels will be regarded as "drop-in" biofuels. According to the Former Undersecretary of the United States Department of Energy, Kristina Johnson, "drop in" biofuels can be defined as fuels produced from various biomass which are compatible with the over $9 trillion energy refinery and gas station infrastructure currently available in the United States (Excerpts from the 4th Annual Cellulosic Biofuel Summit 2009).

Biofuel production involves the use of biomass raw materials extracted from the environment, which are transported to processing sites where the biomass is converted into biofuels. From these processing sites, the produced biofuel is then conveyed to end users through distribution points. It is therefore important to acknowledge the effects of activities and processes at each stage of the life cycle of biofuels on sustainable development – economic feasibility, environmental soundness and social acceptance. A Life Cycle Assessment (LCA) that looks at describing the environmental profile of the whole supply chain of drop-in biofuels is thus important. There are two types of LCA methodologies - the attributional LCA which uses average data for each unit process and the consequential LCA which relies on marginal data for its analysis (Ekvall and Andrae, 2006). Additionally, attributional LCA analysis defines the status quo whilst the consequential LCA measures the impacts through changes in the physical flows.

System dynamics is a mathematical and methodological approach used in understanding complex systems over time. It helps us to better understand the interactions that will go into producing results when interactions between components of a system are simulated. The coupling of system dynamics with a consequential LCA conducted on forest based "drop-in" biofuels in the state of Maine can reveal many intriguing facts that might not be revealed in the forecasting associated with consequential LCA. System dynamics will help us gain qualitative insights into understanding the intricacies of economic, ecological and social issues when they interact with biofuel systems. The results of this study will be presented at the meeting.

(120d) Optimizing the Sustainability of Thermochemical and Biochemical Production of Cellulosic Ethanol

Eric C. D. Tanand Daniel Inman, National Bioenergy Center, National Renewable Energy Laboratory, Golden, CO

To lessen the United States' dependence on the use of imported petroleum for producing various liquid transportation fuels, a number of biofuels development initiatives are underway. Aggressive renewable fuel policies will necessitate unprecedented biomass feedstock production and installation of fuel conversion and distribution infrastructure. Expansion of the cellulosic biofuels industry comes with potentially serious environmental side effects. NREL is actively developing cellulosic biofuel conversion technologies and investigating process optimization to these technologies with the goals of economic and environmental sustainability improvement. The environmental sustainability of the NREL's process designs are assessed on the basis of greenhouse gas emissions, criteria air emissions, water consumption, and energy return on investment. This presentation will highlight the latest state-of-technology (SOT) and process modifications to NREL's biofuel production technology, i.e., thermochemical and biochemical conversion platforms. Sustainability metrics will be assessed and updated based on the latest conversion SOT.

(120e) Advancing Computational Modeling and Analysis for Assessing Sustainability of Large Scale Bioenergy Production

Anthony Halog, Forest Bioproducts Research Initiative/School of Forest Resources, University of Maine, Orono, ME and Nana Bortsie-Aryee, University of Maine, Orono

The study aims to analyze holistically and dynamically the adoption of bio-inspired energy systems, technologies and emergence of its supply chains using trans-disciplinary, integrative and system perspective approaches. An integrated methodological framework has been developed to aid sustainability assessment of emerging systems. This integrated and life cycle computational model for sustainability assessment is based upon the science of complex systems, industrial ecology, systems engineering, ecological economics and biophysical economics.

The interrelationships between system components at various spatial scales can be investigated to produce a sustainable systems behavior. In line with Pareto principle and parsimonious modeling, the interplay of critical criteria, variables, metrics and indicators (i.e. hotspots) for sustainability over time is analyzed. This will help us to describe and possibly prescribe sustainable systems when considering alternative system pathways (which are composed of interacting dynamic sub-systems).

By creating a prototype for dynamic system model for sustainability transition, we can explore visions and scenarios for sustainable development such as the transition from a current (unsustainable) path to sustainable pathways. Through foresighting and backcasting techniques, scenarios can be explored and investigated to create sustainable future. The different scenarios and systems development pathways considered will provide time series results and pattern formations for evolution and development of sustainable systems, which might be useful for prediction. The trends and patterns can facilitate the formulation of strategies and practices of new entrants and incumbents in the pursuit of sustainable entrepreneurship and green industrial transformation.

The integrated framework is currently being applied to engineered systems, particularly bioenergy systems. In the conference, our initial, on-going applications in creating sustainable supply chains for "drop-in Biofuels" will be presented and discussed.

(120f) Jojoba (Jojoba oil and Jojoba rest) As An Alternative Energy Source in Jordan

Zayed Al-Hamamre, Chemical Engineering Department, University of Jordan, Amman, Jordan

The objective of this work is to investigate the energy value of Jordanian jojoba seeds. The oil content in jojoba seeds obtained from jojoba plants cultivated in Jordan is investigated utilizing solvent extraction. The effects of extraction parameters (solvent and time) were studied. The physicochemical properties of the extracted oil (acid value (AV), saponification value (SP), the iodine value (IV), peroxide value (PV), viscosity, density heating value and ash content) were investigated. These properties are of interest when considering biodiesel manufacturing.

The results show that hexane gave the highest oil yield at 58.76% with optimum extraction time of 30 min.

The physicochemical properties were measure as AV=0.712 mgKOH/g, SV = 93.65 mgKOH/g, IV= 43.803 g I2/100g, PV = 1.708 mEq/kg, the viscosity = 20 cp at 40oC. Such low values of these properties and the high energy content of the jojoba oil (44.308 MJ/kg) suggest the possibility of single step base catalyzed process transesterification for biodiesel manufacturing with high yield. Also, the remaining part after oil extraction (jojoba rest) has a heating value of 15. 344 MJ/kg and thus it can be used to replace solid nonrenewable fuels.

(143a) Cationic Polyelectrolytes Decrease Gelatinization Temperature of Cornstarch

Sandeep Mora, Chemical and Biomolecular Engineering, Georgia Institute of Technology, Atlanta, GA, Kendra E. Maxwell, School of Chemical & Biomolecular Engineering Georgia Tech, Georgia Institute of Technology, Atlanta, GA and Sujit Banerjee, School of Chemical and Biomolecular Engineering, Georgia Institute of Technology, Atlanta, GA

Cationic polyelectrolytes such as c-PAM and poly-DADMAC decrease the gelatinization temperature of cornstarch granules. Enzymatic hydrolysis of cornstarch is, therefore, more efficient in the presence of the polymers. Viscosity profiles of various combinations of the cornstarch/enzyme/polymer system will be presented as will corresponding glucose and Brix values. A mechanism for the effect of the polymer will be proposed and discussed with reference to previous work on polyelectrolyte assisted hydrolysis of bleached cellulosic fiber.

(143b) Cloud Point Studies for Diesel and Jet Fuel Bio-Derived Fuels

Anne Lown, Lars Peereboom, Dennis J. Miller and Carl T. Lira, Chemical Engineering and Materials Science, Michigan State University, East Lansing, MI

Alternatives to petroleum based fuels are essential for the future transportation industry. For a fuel to be a viable replacement for a petroleum fuel it must meet a number of specifications. This work considers the compatibility of potential biofuels with diesel and jet fuel. This research focuses on low temperature properties.

Cloud point measurements are presented for additives in two test diesel blends and jet fuel. Additive functionalities studied include, diesters, esters, ketones, ethers, and alkanes. Diesters, as a class, appear to create liquid-liquid immiscibility and a cloud point higher than that of either the base fuel or the pure diester. Esters decrease the cloud point of diesels fuels, as do most ketones. Ethers and alkanes respectively decrease the cloud point even more. These results are consistent with the precipitation of a component of the petroleum fuel, rather than an additive, and the decrease in cloud point is due to a dilution effect. For some ketones, a different effect is observed. Based on the pure component melting temperature of the ketone, the compounds studied cause decreases or increases.

Long chain ester components were also tested. These additives have higher cetane numbers than other potential fuel components, and are therefore more desirable as a diesel fuel component. Ketones and ethers of similar lengths were also tested, and have the same behaviors as described above. The behavior of potential fuel additives seems to be more related to the oxygen containing functional group rather than extent of branching or size.

Experimental results will be interpreted with thermodynamic models using modified surrogates.

(143c) Process Development and Demonstration of Biomass to Liquid (BTL) Fuels Via Gasification and Catalytic Conversion

Qiangu Yan, Department of Agricultural & Biological Engineering, Mississippi State University, Mississippi State University, MS; Dave C. Swalm School of Chemical Engineering, Mississippi State University, Mississippi State University, MS and Fei Yu, Department of Agricultural & Biological Engineering, Mississippi State University, Mississippi State, MS

Biomass like wood chips, switchgrass and other plant residues are first converted to syngas through gasification process using air, oxygen or steam. A downdraft gasifier is performed for syngas production in Mississippi State. The syngas from the gasifier contains about 20% H_2, 20% CO, 11% CO_2, and 49% N_2. In addition to the gases listed above, the bio-syngas also contains some tars, ammonia, H_2S, and particulates as impurities. Bio-syngas has to be polished prior to CO hydrogenation to liquid fuels. The purpose of the work is to catalytic upgrade cleaning syngas to higher hydrocarbons and value-added chemicals. A series of catalysts have been developed to produce gasoline, diesel, jet fuels, olefins, higher alcohols and esters. Current work successfully demonstrates the process of biomass to liquid (BTL) fuels via gasification and catalytic conversion.

(143d) Global Optimization of Oil Production in Microalgae Bioreactor

Javad Abdollahiand Stevan Dubljevic, Chemical and Materials Engineering, University of Alberta, Edmonton, AB, Canada

Sustainability and renewability of bio-fuels make them suitable replacements for fossil fuels, however because of slow production rate and high production costs, theyare not economically comparable with traditional fuel sources. Bio-fuel production fromphototrophic and heterotrophic microalgae is a novel method of oil production that isgoing to replace bio-fuel production from crops. In this method there is no need forvast land, high amounts of water and in the case of heterotrophic microalgae there is noneed for sunlight. Furthermore, increasing oil production rate can ensure the economicfeasibility of biodiesel production.

This work is built on an existing heterotrophic microalgae fed-batch model, see [1], whichdescribes the algae growth and oil production. The fed-batch microalgae reactor is fedby two nutrients, glucose as carbon source and glycine as nitrogen source. In this work,maximum possible oil production in the shortest time interval is calculatedby using a global optimization technique. The highly nonlinear rate functions, timevarying parameters and uncertainties in the modeling are taken into accountwithin the optimization framework.

Reference:

[1] H. De la Hoz Siegler, A. Ben-Zvi, R.E. Burrell, W.C. McCa rey, The dynamics of heterotrophicalgal cultures, Bioresource Technology 102 (2011) 5764-5774

Sustainable Biodiesel Production from Waste Cooking Oil and Animal Fats under Supercritical Methanol

AICHE 2012 Spring Meeting, Houston, Texas

Chee Kai Tan, Maoqi Feng, Laura Harris, and Cary Wong
Chemical Engineering Department, Chemistry and Chemical Engineering Division, Southwest Research Institute, San Antonio, Texas

Abstract

In the conventional process for transesterification of vegetable oils and animal fats to biodiesel, free fatty acids and water always present a problem and result in a low conversion. In this study, triglycerides, animal fats, and waste cooking oil were used as feedstock for transesterification under supercritical methanol conditions. For glyceride tributyrate, the conversion was 100% under SCM conditions. For waste cooking oil, the fatty acid methyl ester (FAME) yield was >90% at the following reaction conditions: reaction temperature = 300°C and 250°C, reaction pressures = 2000 psi, 2500 psi, and 3000 psi, residence time = 18 minutes, and molar ratio of methanol to oil = 40:1. However, when the residence time decreased to 7 minutes, at 250 °C and 2500 psi, the transesterifcation conversion dropped to 86 %. Using a heterogeneous catalyst, ETS-10, the transesterification reaction conditions were lowered to subcritical levels, and a conversion of over 99% was achieved with a residence time of 7 minutes. Because of the short residence times, reduced waste, and high triglyceride conversions, a biodiesel process employing supercritical methanol may present significant economic advantages over the conventional manufacturing process.

Introduction

As the availability of petroleum-derived diesel fuel slowly decreases, the interest in alternative fuels grows. Biodiesel refers to a vegetable oil or animal fat based fuel consisting of long-chain fatty acid alkyl esters (FAMEs). It is a renewable alternative fuel and has been successfully derived from lipids (vegetable oil or animal fat) and an alcohol as shown in Equation 1.

CH_2-OCO-R						CH_2-OH	
|						|	
CH-OCO-R	+	3R'OH	—>	3RCOOR'	+	CH-OH	(1)
|		Alcohol		Alkyl Ester		|	
CH_2-OCO-R						CH_2-OH	
Triglyceride						Glycerol (byproduct)	

R = fatty acid chain
R' = CH_3 : fatty acid methyl esters
R' = CH_2CH_3 : fatty acid ethyl esters

Biodiesel is becoming well-known as an environmentally friendly fuel due to its nontoxic and biodegradable characteristics. It has reduced engine emissions of sulfur oxides, carbon dioxide, particulate matter, and aromatic hydrocarbons compared to conventional diesel fuels. Due to federal legislation to promote energy independence (EISA) and the consequent mandates issued by the EPA (in RFS2) requiring the increased production of transportation biofuels, biodiesel has received considerable attention: it appears to be one of the most promising of the biofuels for meeting the production volume requirements.

Research into the use of transesterified sunflower oil as diesel fuel was initiated in South Africa in 1979. By 1983, the process for producing fuel-quality, engine-tested biodiesel was completed and published internationally [1]. Biodiesel is commercially produced in batch or continuous stirred tank reactors at temperatures ranging from 60 to 200°C using homogeneous base catalysts with edible vegetable oils such as palm oil, corn oil, sunflower oil, and soybean oil. The fatty acid triglycerides are converted to the respective alkyl ester (biodiesel) and glycerol by acid-catalyzed [2-4] or base-catalyzed [5-11] transesterification in the presence of an excess of methanol or ethanol.

However, the use of homogeneous base catalysts requires neutralization and separation from the final reaction mixture and results in higher solvent consumption. Thus, these additional production costs make the product biodiesel not cost competitive with fossil diesel. Heterogeneous solid base catalysts, able to catalyze the transesterification of triglycerides to alkyl eaters, could solve these problems. The solid catalysts can be separated from the biodiesel without using solvent and they can be easily regenerated [12-17].

Transesterification of oils to alkyl esters in supercritical methanol without a catalyst was developed by Saka and Kusdiana [18-20] and offered a promising future for industrial applications. Supercritical fluids are fluids above their critical temperature and pressure. The supercritical fluid behaves gas-like considering diffusivities but also liquid-like with respect to viscosities. As shown in Figure 1, the critical temperature and critical pressure of methanol are 240°C and 1140 psia, respectively [21].

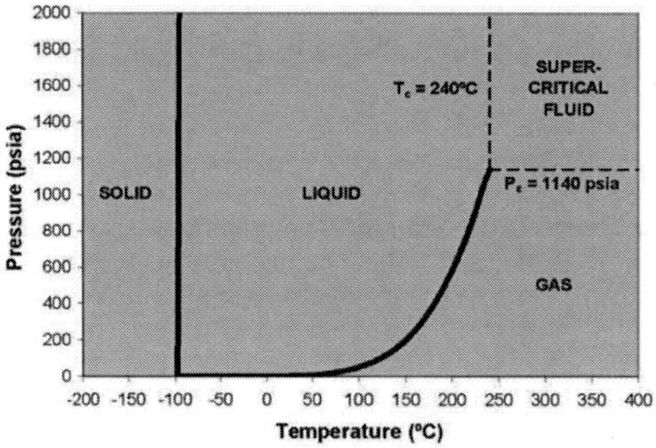

Figure 1. Supercritical methanol phase diagram.

Nonpolar triglycerides can be well solvated with supercritical methanol to form a single phase oil/alcohol mixture. In addition, the conversion rate of oil into esters increases dramatically in supercritical alcohols, the reaction does not require a catalyst, and it is not affected in the presence of water and free fatty acids [22-28]. In a supercritical alcohol, the transesterification reaction was nearly completed in less than 15 minutes for both edible and non-edible oils [26]. Using the supercritical methanol method for conversion waste cooking oil at 1450 psi and a temperature of 300°C, the yield of biodiesel was shown to increase to 50-65% in the first 15 minutes [27]. Recently, a continuous production process for vegetable oil conversion to biodiesel through the supercritical methanol (SCM) has been reported [29-31]. However, continuous SCM processing of waste cooking oil and animal fats has not been studied. This paper summarized our study of such a process.

Experimental Section

1) Reactor System: A flowsheet of the continuous transesterification fixed-bed reactor system with supercritical methanol is shown in Figure 2. Figure 3 [32-36] is a photo of the system as built in the Chemical Engineering Center at Southwest Research Institute.

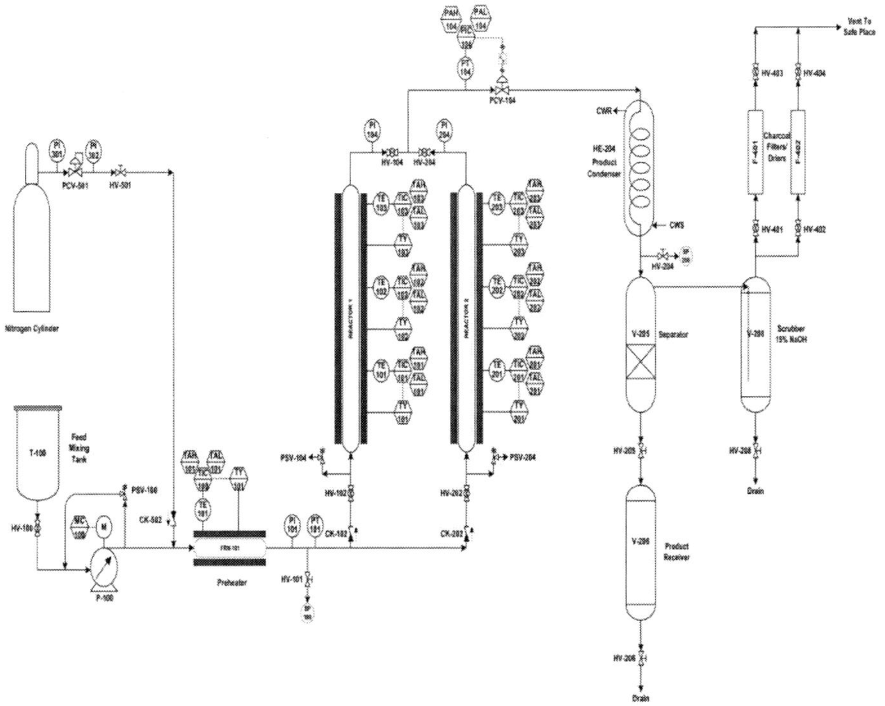

Figure 2. Process Flow Diagram for biodiesel production with supercritical methanol.

Figure 3. Continuous transesterification bench unit at Southwest Research Institute.

The reactor was constructed from a 3/8" OD x 4' long stainless steel tube with a wall thickness of 0.049 inch, giving a volume of 47.4 mL. The reactor was built with two thermowells welded into the reactor tube and located one third of the way from each end of the reactor. Thermocouples inserted into the thermowells monitored the internal temperature of the chemicals as well as the catalyst temperature. The stainless steel reactor was wrapped with heat tape and installed inside an aluminum enclosure insulated with a calcium silicate material.

The pressure within the reactor system was recorded with a pressure transducer and regulated with a pressure control valve, both located downstream of the reactor. At the outlet of the pressure controller, the biofuels, the bio-based byproduct, glycerin, and excess methanol were then fed into a condenser and collected in a graduated glass receiver. Uncondensed gases were captured within an activated-carbon scrubber. For safety, pressure relief valves were installed along the reactor and the transfer lines. The process instrumentation was automatically controlled with an i-Fix program. The software allowed for data trending, automatic controls, safety alarms and fail-safe process shutdown.

A 4L feed preparation vessel mixed the reagents, feedstock and methanol, using as overhead mechanical stirrer. The feed vessel was located on a weigh scale so that the delivery of reagents to the reactor could be measured for material balances. The reagents were pumped to the reactor using a high-pressure Eldex metering pump (Model B-100-S, flow rate from 0.2 – 8.0 mL/min @5000 psig). The inlet of the reactor was equipped with a pressure gauge and transducer to monitor the liquid feed pressure. A pressure safety valve was installed at the inlet of the reactor and set at 4,000 psig.

2) Feedstock Preparation

Three feedstocks were prepared: methanol/tributyrin (40:1 molar ratio), methanol/waste cooking oil (40:1 molar ratio), and methanol/pork fat (>200:1 molar ratio). The first feedstock, consisting of methanol and tributyrin, was a single phase, clear solution. No filtration or other purification was needed before combining the compounds or running them through the system.

The waste cooking oil (WCO) was filtered first through a 10-micron filter followed by a 0.7-micron filter to remove suspended impurities that could potentially plug the system, particularly the catalyst pores. After combining this filtered WCO with the appropriate amount of methanol based on stoichiometry, the solution was biphasic and thus had to be constantly mixed to maintain a homogeneous consistency while being pumped into the system.

The pork fat was in a semisolid state before use, and it therefore had to be heated prior to filtering. The fat was filtered through a 10-micron filter before it was combined with methanol in a 200:1 molar ratio of methanol to pork fat. This molar ratio was increased from the typical 40:1 molar ratio in order to solubilize more of the pork fat in the methanol. The feedstock was biphasic and was mixed constantly.

3) Catalysts Preparation

Microporous Titanosilicate ETS-10 Catalyst [37]

184.1 g of water and 182.5 g of sodium silicate solution (10.6 wt % Na2O, 26.5 wt % SiO_2) were added to a Teflon beaker and the mixture was stirred. While stirring, 63.2 g NaCl, 12.3 g KF, and 12.3 g TiO_2 were added into the silicate solution producing a translucent thick gel. After the solution was stirred vigorously for 30 minutes at room temperature, the mixture contained in the Teflon beaker was placed into a Parr autoclave and heated to 200 °C for 42 hours. The gel had the composition $5.3SiO_2:1.0TiO_2:1.6Na_2O:7.2NaCl:1.0KF$ or $KCl:110H_2O$, and the pH was around 10.5. The samples obtained were denoted as microporous titanosilicate ETS-10.

4) Analytical Methods

The biodiesel was analyzed with a Hewlett Packard GC Model HP 6890 interfaced with an Agilent 5973 Network Mass Selective Detector (Figure 9). The analytical column was a DB-23 fused-silica capillary column (0.25mm ID x 60m length) with a 0.15μm film thickness silicone-coated fused-silica capillary column. This narrow-bore capillary column was directly connected into the mass spectrometer inlet.

Direct measurement of triglycerides present in the converted product samples were accomplished using high performance liquid chromatography (HPLC) coupled with an evaporative light scattering detector

(ELSD). By analyzing feedstock and product samples, a triglyceride conversion rate was determined. The analytical method was developed on a Waters HPLC system model LC03, utilizing a 600E pump, 717 autosampler and an Alltech 2000 ELSD. A Supleco brand analytical column (part number Supelcosil LC-18-DB, 150 mm x 4.6 mm.) was chosen for the resolution of the triglycerides present in the samples.

Results and Discussion

1) Baseline: Transesterification of tributyrin under supercritical methanol conditions

Tributyrin was used as benchmark compound to guide the experiment. Table 1 shows the results of the GC/MS analyses conducted on the tributyrin samples. The data indicate that at temperature below supercritical condition of methanol (SCM), there was no conversion of tributyrin. At temperature above SCM (240 °C, 1140 psi), tributyrin was completely transesterified to butyric acid methyl ester with methanol.

Table 1. GC-MS Results and Percent Conversions

Run ID:	**6A**	**6B**	**6C**	**6D**	**6E**	**6F**
Feedstock	Tributyrin	Tributyrin	Tributyrin	Tributyrin	Tributyrin	Tributyrin
Pressure Stpt (psig)	2000	2500	3000	2000	2500	3000
Temp Stpt (°C)	200	200	200	300	300	300
Initial Tributyrin Concentration (mg/L)	158147	158147	158147	158147	158147	158147
Final Tributyrin Concentration (mg/L)	169000	180000	150000	0	0	0
Percent Conversion (%)	-6.86	-13.82	5.16	100	100	100

2) Transesterification of WCO under supercritical methanol conditions

The results of transesterification of WCO with SCM (> 240 °C/1140 psig) are tabulated in Table 2. The results show that high conversions of triglyceride into fatty acid methyl esters (FAMEs) with SCM are consistently achievable on a continuous basis with reaction conditions in the supercritical region for methanol: temperature ranges 250 – 325 °C and pressure ranges 1500 – 3000 psig at a feedstock flow rate of 2.5 mL/min.

Table 2. Waste Cooking Oil (WCO) Biodiesel Conversion with Supercritical Methanol (SCM).

Sample ID	**Temp. °C**	**Pressure, psig**	**Flowrate, mL/min.**	**Residence Time, min.**	**TG (mg/mL) Concentration**	**Triglyceride Conversion (%)**
R8145-8A-033	300	1500	2.78	17.16	33	**88.2**
R8145-8B-035	300	2500	2.6	18.35	0.34	**99.9**
R8145-9A-037	250	1500	2.53	18.85	7	**97.5**
R8145-9B-039	250	2500	2.2	21.68	0.93	**99.7**
R8145-9C-041	250	3000	2.4	19.88	0.56	**99.8**
R8145-9D-043	300	3000	2.58	18.49	0.56	**99.8**
R8145-10A-045	325	1500	2.3	20.74	0.54	**99.8**
R8145-10B-046	325	2500	2.55	18.71	0.58	**99.8**
R8145-11A-049	325	1500	2.4	19.88	8.2	**97.1**
R8145-11B-050	325	2500	2.67	17.87	0.59	**99.8**
R8145-25A-076	250	2500	4.7	10.15	17	**93.9**
R8145-25B-077	250	2500	5.57	8.56	26	**90.7**
R8145-25C-078	250	2500	7	6.81	38	**86.4**

R8145-28A-083	250	2000	2.83	16.86	3.6	**98.7**
R8145-28B-084	250	1700	2.86	16.68	2.2	**99.2**
R8145-28C-085	220	2200	3	15.9	18	**93.6**

However, as shown in figure 4, the conversion dropped to 86.4% when the flow rate increased from 2.2 mL/min (residence time 21.68 min) to 6.81 mL/min (residence time 6.96 min), which indicates that triglyceride conversion is directly dependent on the contact time of the triglyceride with methanol (MeOH).

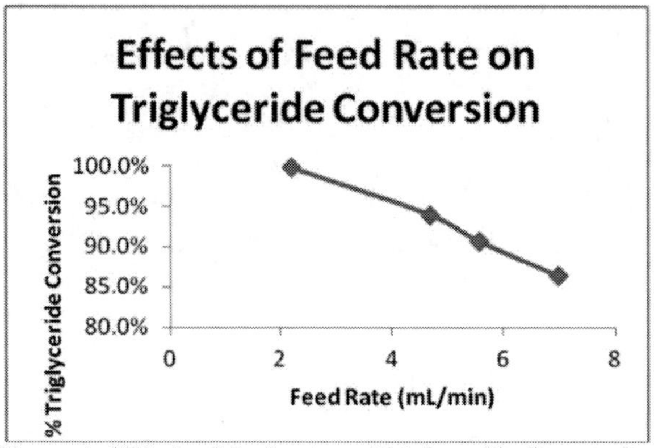

Figure 4. Effects of feed rate on Triglyceride Conversion

3) Transesterification of animal fat under supercritical methanol conditions

Table 3 shows the transesterification results of animal fat with SCM where high conversions were achieved with temperature in the range of 250 °C – 300 °C incorporated with pressure in the range of 2000 psig – 3000 psig. The Eldex pump was not able to perform with higher concentrations of pork fat due to associated plugging problems of the highly viscous feedstock.

Table 3. Pork Fat Biodiesel Conversion with Supercritical Methanol (SCM).

Sample ID	Temp. °C	Pressure, psig	Flowrate, mL/min.	Residence Time, min.	TG (mg/mL) Concentration	% Conversion
R8145-13A-051	300	2000	2.93	16.28	2	**96.4**
R8145-13B-053	300	2500	2.8	17.04	0.75	**98.6**
R8145-14A-055	300	3000	2.6	18.35	0.05	**99.9**
R8145-14B-056	250	2500	2.87	16.62	0.04	**99.9**
R8145-14C-057	250	3000	2.5	19.08	0.16	**99.7**

4) Effect of catalyst at low reaction residence times

At 250 °C/2500 psig/flowrate 6.8 mL/min., the transesterification of WCO into FAMES was low as cited in Table 2 (86.4% conversion). These reaction parameters were selected to the effect of the presence of a solid catalys by comparing the WCO transesterification activity with SCM/catalyst to SCM without catalyst. The reactor was packed with ETS-10 Catalyst and operated for twenty-four hours with SCM. Figure 5 shows that the conversion efficiency with catalytic remained active throughout 24 hours of experiment, conversion is consistently achieving >99% with a residence time of 7 minutes.

Under the same conditions, 250 °C/2500 psig/flowrate 6.8 mL/min., the transesterification of WCO to FAMEs with SCM dropped in the range of 36.5% -58.6 % in the 12 hour, nonstop experiment as shown in Figure 5. The experiment demonstrated that solid catalyst is needed at sub-supercritical conditions.

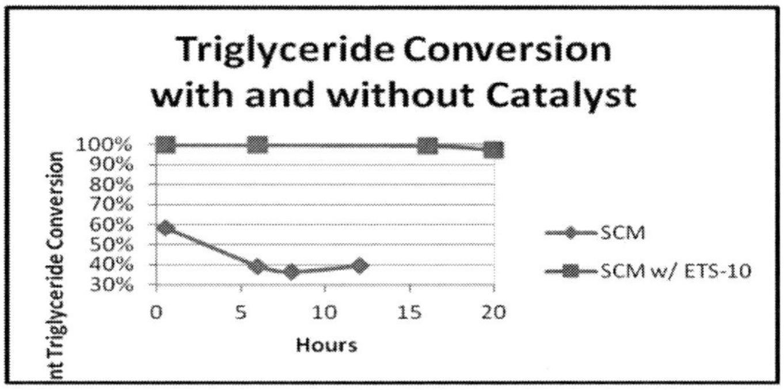

Figure 5. The effect of a solid catalyst on triglyceride conversion under SCM conditions.

Summary

Biodiesel production with continuous SCM processing of waste cooking oil and animal fats has been successfully demonstrated in lab scale in our study. Presence of a solid catalyst dramatically improves triglyceride conversion at low reaction residence times. Development of a continuous SCM biodiesel process on a pilot-plant scale is in progress at Southwest Research Institute.

Acknowledgement

The authors would like to take this opportunity to thank the Internal Research Program, Southwest Research Institute, for financial support.

References

1. K.J. Harrington and D.V. Catherine, A comparison of conventional and in-situ methods of transesterification of seed oil from a series of sunflower cultivars, J. Am. Oil Chem. Soc. 62 (1985), pp. 1009-1013.
2. J. Otera, Transesterification, Chem. Rev. 93 (1993), pp. 1449-1970.
3. B. Davies and G.V. Jeffreys, Continuous transesterification of ethyl alcohol and butyl acetate in a sieve plate column: II. Batch reaction kinetic studies, Trans. Inst. Chem. Eng. 51 (1973), pp. 271-274.
4. B. Freedman, R.O. Butterfield and E.H. Pryde, Tranesterification kinetics of soybean oil, J. Am. Oil Chem. Soc. 63(1986), pp. 1375-1380.
5. J. Schmidt, D. Reusch, K. Elgeti and R. Schomacker, Kinetics of ethanol and butyl acetate-a model system for reactive rectification, Chem Ing. Tech.71 (1999), pp. 704-708.
6. P. Felizardo, M.J.N. Correia, I. Raposo, J.F. Mendes, R. Berkemeier and J.M. Bordado, Production of biodiesel from waste frying oils, Waste Management, Vol. 26, Issue 5, 2006, pp. 487-494.
7. A.N. Phan and T.M. Phan, Biodiesel production from waste cooking oils, Fuel, Vol. 87, Issues 17-18, 2008, pp. 3490-3496.
8. S. Gryglewicz, Alkaline-earth metal compounds as alcoholysis catalysts for eater oils synthesis, Appl. Catal. A 192 (2000), pp. 23-28.

9. A. Corma, S. Iborra, S. Miquel and J. Primo, Catalysts for the production of fine chemicals, J. Catal. 173 (1998), pp. 315-321.

10. D. Damoko and M. Cheryon, Kinetics of palm oil transesterification in a batch reactor, J. Am. Oil Chem. Soc. 77 (2000), pp. 1263-1267.

11. K. Komers, J. Machek, R. Stloukal, Biodiesel from rapeseed oil and KOH 2. Composition of solution of KOH in methanol as reaction partner of oil, Eur. J. Lipid Sci. Technol. 103 (2001), pp. 359-462.

12. G. Guan, K. Kusakabe and S. Yamasaki, Tri-potassium phosphate as a solid catalyst for biodiesel production from waste cooking oil, Fuel Processing Technology, Vol. 90, Issue 4 (2009), PP. 520-524.

13. K.G. Georgogianni, A.P. Katsoulidis, P.J. Pomonis and M.G. Kontominas, Transesterfication of soybean frying oil to biodiesel using heterogeneous catalysts, Fuel Processing Technology, Vol. 90 Issue 5 (2009), pp. 671-676.

14. C. Reddy, V. Reddy, R. Oshel, and J.G. Verkade, Room-Temperature Conversion of Soybean Oil and Poultry Fat to Biodiesel Catalyzed by Nanocrystalline Calcium Oxide, Energy & Fuels 20 (2006), pp. 1310-1314.

15. C.V. McNeff, L.C. McNeff, B. Yan, D.T. Nowlan, M. Rasmussen, A.E. Gyberg, B.J. Krohn, R.L. Fedie, T.R. Hoye, A continuous catalytic system for biodiesel production, Appied Catalysis A: General 343 (2008), pp. 39-48.

16. Y.M. Park, D.W. Lee, D.K. Kim, J.S. Lee, K.Y. Lee, The heterogeneous catalyst system for the continuous conversion of free fatty acids in used vegetable oils for the production of biodiesel, Catalysis Today, 131 (2008), pp. 238-243.

17. N.S. Babu, R. Sree, P.S. Sai Prasad, and N. Lingaiah, Room temperature transesterification of edible and non-edible oils using a heterogeneous strong basic Mg/La catalyst, Energy & Fuels, 22 (2008), pp. 1965-1971.

18. D. Kusdiana, S. Saka, Effects of water on biodiesel fuel production by supercritical treatment, Bioresour. Technol. 91 (2004), pp. 289-295.

19. D. Kusdiana and S. Saka, Kinetics of transesterification in rapeseed oil to biodiesel fuels as treated in supercritical methanol, Fuel 80a (2001), pp. 693-698.

20. D. Kusdiana and S. Saka, Methyl esterification of free fatty acids of rapeseed oil as treated in supercritical methanol, Chem. Eng. Jpn., 34 b (2001), pp. 383-387.

21. R.E. Babcock, E.C. Clausen, M. Popp, W. B. Schulte, Yield Characteristics of Biodiesel Produced from chicken fat-tall oil blended feedstocks, http://ww2.mackblakwell.org/web/research/all_research-projects, 02/09/12.

22. D. Kusdiana and S. Saka, Effects of water on biodiesel fuel production by supercritical methanol treatment, Bioresource Technology, 91 (2004), pp. 289-295.

23. Y. Warabi, D. Kusdiana, S. Saka, Reactivity of triglycerides and fatty acids of rapeseed oil in supercritical alcohols, Bioresource Technol. 91 (2004), pp. 283-287.

24. W.L. Coa, H.W. Han and J.C. Zhang, Preparation of biodiesel from soybean oil using supercritical methanol and co-solvent, Fuel 84 (2005), pp. 347-351.

25. V. Rathore, G. Madras, Syntheis of biodiesel from edible and non-edible oils in supercritical alcohols and enzymatic synthesis in supercritical carbon dioxide, Fuel 86 (2007), pp. 2650-2659.

26. A. Demirbas, Biodiesel from sunflower oil in supercritical methanol with calcium oxide, Energy Conv. Manag. 48 (2007), pp. 937-941.

27. A. Isigigur, F. Karaosmonoglu, H.A. Aksoy, Methyl ester from safflower seed oil of Turkish origin as a biofuel for diesel engines, Appl. Biochem. Biotechnol., 45/46 (1994), pp. 103-12.

28. A. Demirbas, Biodiesel fuels from vegetable oils via catalytic and non-catalytic supercritical alcohol transesterifications and other methods: a survey, 44 (2003), pp. 2093-2109.

29. H. He, T. Wang, S. Zhu, Continuous production of biodiesel fuel from vegetable oil using supercritical methanol process, Fuel 86 (2007), pp. 442-447.

30. K. Bunyakiat, S. Makmee, R. Sawangkeaw, S. Ngamprasertsith, Continuous production of biodiesel via transesterfication from vegetable oils in supercritical methanol, Energy & Fuels 20 (2006), pp. 812-817.

31. H.Y. He, T. Wang, S. Zhu, Han, Continuous production of biodiesel fuel from vegetable oil using supercritical methanol process, Fuel 86 (2007), pp. 442-447.

32. Tan, C.K., Feng, M., Novel biodiesel Production Technology, National Biodiesel Technical Workshop, November 1-2, 2011, Kansas City, Kansas.

33. Tan, C.K., Feng, M., Continuous Biodiesel Production from Waste Cooking Oil and Animal Fats with Supercritical Methanol Conditions, USDA-RD State Director Conference, March 30, 2011, San Antonio, Texas.

34. Tan, C.K., Feng, M., Continuous Biodiesel Production from Waste Cooking Oil and Animal Fats under Sub/Supercritical Methanol Conditions, 2011 AIChE Spring Conference, March 13-17, Chicago.

35. Tan, C.K., Feng, M., Cooking Waste to Biofuels, USDA 9007 Renewable Energy for America Program - Workshop, January 26, 2011, Uvalde, Texas.

36. Tan, C.K., Feng, M., Transesterification of triglyceride tributyrate and waste cooking oil with supercritical methanol, 2010 ACS Winter Conference, November 30-December 4, 2010, New Orleans.

37. X. Yang, J.-L. Paillaud, H.F.W.J. van Breukelen, H. Kessler, E. Duprey, "Synthesis of microporous titanisilicxate ETS-10 with TiF4 or TiO2", Microporous and Mesoporous Materials 46 (2001) 1-11.

Unlimited Power from Continental Drift, 284207

Donald I. Garnett

Overview: Given a place to stand and a long enough lever Archimedes (1) said he could move the world. Donald I. Garnett (2, 3) observes that anchoring one continent to another with a cable will let the world's continental drift (4, 5) move you, namely: it will provide all the energy needed to power the entire worlds electrical (6) and other needs. To capture the energy of the moving continents it will be especially convenient to catenary a cable across a fault line and let the mass of the cable plus any added weights desired be lifted in elevation against gravity by the moving apart of the continents. Recovery of that energy can be accomplished using an electric generator driven by the pull of the cable in restoring the cables initial position.

Restraining the movement of a continent the size of North America in theory could generate 10^10 times the electrical usage of the United States. In principle its' restraint might stop or prevent earthquakes in California, Oregon and Washington.

The magnitude of the energy available exceeds that of tearing down the Rocky Mountains (in cable cars attached to electric motors) and dropping the rocks into the Pacific Ocean. Had the erosion by Nature of the Rocky Mountains not occurred it would be closer to equivalent but still inferior?

While this method may be applicable in many places at continental plate boundaries it is easily visualized for the Rift Valley in Africa. Places where a cable would acquire tension by continental plate movement being preferred. Movement at the face of any type fault however can be harnessed. Lateral, vertical, or oblique plate movement energy recovery can be accommodated by inclusion of a spring in the cable. Multiple echelon faults energy similarly can be harnessed using a network of spring loaded cables.

Methods: The energy associated with the drift of a continent is its ½ *mass * velocity^2 from a given anchor point. In this discussion the anchor point is another continental plate reachable by a cable. Across a fault where the movement is "apart" is especially preferred. The velocity is low but the mass is extremely large. The usual ½ mass times velocity squared is also (½*mass*velocity/410.3Kw)*velocity, or Power*velocity. With mass of a continent being say 10^22 lbs and velocity of 2.5 cm/year the power available is 10^15Kw or 10^10 times the total U. S. Electrical power.

The anchor points need not be capable of restraining the continent but may be any amount consistent with the intended power output. The small movement of the continent and the preferred relatively high velocity of the cable driving the generator begs for a velocity enhancement like a block and tackle arrangement or a geared reducer run backwards. In any event a 1000 fold block and tackle on top of another (for leverage per Archimedes) , etc., etc., will likely be in the final design. Other methods of energy recovery are possible, including:

(1) Spring compression and release,
(2) Hydraulic lifting of water for use in turbines on release.
(3) Gas compression and release.

Historically of course part of the energy of the continents movement has raised mountains and provides the force multiplier and accumulator of millions of years of potential energy generated by the continents movement. Recovery of that potential energy is available by tearing down the mountains, for example in cable cars connected to electric generators and dumping the rocks into the oceans continental trench. The Rocky Mountains, the Alps, the Andes and the Himalayas for example are available at present. Over time both energy and material (mountains) are conserved

Results: The result of using continental drift for essentially unlimited power for the human race is probably unfathomable, but nearly infinitely describable, including: No green house gasesThe hydrogen economy.

1. The elimination of nuclear power plants.
2. Electric cars with no emissions and other transportations.
3. Electric aero planes.
4. Unlimited agriculture from electric agro-lights. Production can become volumetric vice area limited, further postponing Malthus effects.
5. The largest possible scale for energy production ergo least cost.
6. Waste disposal reduced to near zero.
7. Industrial production with only carbon, water and electricity.
8. CO_2 impoundment or removal for its carbon content for global climate control.
9. Oxygen supplementation of the atmosphere for enhanced ozone protection of earth.
10. Population explosion from all of the foregoing.
11. Focused power into space.

Conclusions: Unlimited power from continental drift for the human race contains the essentials for future growth and sustainability of the race.

Some practical considerations:

While the movement of the continents has the potential power to overwhelmingly provide the needs of mankind, collecting it presents some practical problems that need to be overcome by Engineering.

Foremost among them is the issue of scale. If one supposes the power plants will be built say 100 years in the future, and the magnitude of the facility will be such as to satisfy the needs of humanity then it is safe to say the Scale will be at least 100 fold larger than the largest presently known facility. The three gorges dam and power plant in China may be a useful reference for this visualization. At 100 fold that scale the price of electricity/kwh will be down (in constant dollars) by 100 fold, assuming no increase in process complexity. Basically that means hydro electric power with a similar hydrostatic head and volumetric flow.

Major Parameters of the TGHP units and abroad power plants

Parameters	**Three Gorges Project** CHINA	**Grand Coulee** U.S.A.	**Itaipu** Brazil/Paraguay	**Guri** Venezuela	**Krasnoyarsk** Russia

Turbine					
Maximum Head(m)	113.0	108.2	126.7	146	100.5
Rated Head(m)	80.6	86.9	112.9	130	93
Minimum Head(m)	71.0(61.0)	67.0	82.9	111	76
Rated Output (MW)	710	612/716	715	610	508
Maximum Output(MW)	852	827	740/808	730	508(505)
Rated Spinning Speed(r/min)	75	85.7	90.9/92.3	112.5	93.8
Runner Diameter(m)	9.525(9.800)	9.86/9.22	8.45	7.163	7.5
Generator					
Rated Capacity(MW)	778	718	823.6/737.0	700	500
Maximum Capacity(MW)	840	710/825.6	823.6/766	805	
Frequency(Hz)	50	60	50/60	60	50
Cooling	Water Cooling of Statar	Water Cooling of Statar	Water Cooling of Statar	Air Cooling	Water Cooling of Statar
Rated Voltage(KV)	20	15	18±5%	18	15.75
Insulation Levels	F	B	F	B	
Thrust Load(t)	4,050(4,100)	4,700	4,056&4,400	2,667	
Inner Diameter of Stator(m)	18,500(18,800)	18,790	16,000	13,650	16,100

Preferred Conditions:

While this article started off with cables, multiple cables, springs, cable velocity enhancers and the passage of large amounts of time, the scale, and volumetric constraints imposed by costs lead to preferred conditions as follows and utilizes the energy accumulated over millions of years (not just a one-time event).

- Use gravity vice cables and springs.
- Use of water as the motive force for electric turbines or direct mechanical connection to the delivery conveyer. That means on the coast, close to continental trenches.
- Hydrostatic head >200 ft., so use the continental trench at 10-20,000 ft., a scale advantage of 50-100 fold.
- Transport of the mass of the continent being sent to the bottom of the trench should be as a fluid itself. Examples of two such masses are (1) the sands of the Sahara Desert, and (2) the mud volcanoes near the Mariana trench and elsewhere. An additional alternative could be molten lava, for example in Hawaii where the ocean trenches are nearby. Using
- the lava that has been cooled by Nature but is in danger of sliding off the island and creating tsunamis could be considered but I will leave that and other alternatives to others.

- This alternative suggests that the rock should be pulverized and transported as a fluid to the sea. The Western Sahara is an example of where nature over time has pulverized mountains and the wind (fluid) has transported it to the sea shore (and beyond with a gradient of about 2% out to sea). A 100 mile pipeline will yield a head for the power plant of 2 miles (10560 ft.) vs. say the 3 gorges head of 200 ft., for a scale advantage of at least 50 fold.
- The water for the electric turbines and used to transport the solid mass to the trench, may well have to be pumped from the sea to the mine, but considering the power generated by a solid of 2.5 specific gravity, dropping 10,000+ ft. Compared to the water at 1.03 specific gravity being pumped a small fraction of that height, should not be a limitation.
- The transport of the solids to the bottom of the trench could be done by basically a bucket elevator e.g., rail cars (at least 100 times the capacity of the present cars, i.e. a 5 fold scale increase in 3 dimensions or 125 fold) with return. Such delivery and return may preferably be encased in piping.
- As mentioned initially some form of **velocity enhancement** (more leverage per Archimedes) should be included, and by moving continental shelf mass down to the continental trench at say 45 miles/hour is a velocity enhancement, over the 2.5 cm/year movement of the continent, of 2*10^10. Combined with the head advantage of 50 fold yields an overall advantage of 10^12 over collecting the energy at a fault line (2, 11).
- Transfer of power from the rail car chain to the electric turbines could be by mechanical gearing or by suitable sealing around the rail cars and a larger diameter return pipe conduit (up to 2.5/1.0 times the area of the delivery pipe and its discharge and specified elevation above sea level) to supply electric turbines. One could simply use the volume of water returned in the emptied rail cars by extending the return pipe in elevation of say 1000 to 5000 feet above sea level. Other alternatives may include magnetic levitation of the buckets (rail cars) and one supposes magnetic braking could recover the electrical energy without need for a separate turbine/generator/power station.
- Mining of the sand deposits on shore may not be necessary. Hydraulic dredging of offshore deposits as the source (mine) and sand slurry transfer to the pipeline system for delivery to the continental trench may be preferred. Combined with magnetic levitation and magnetic braking for electricity generation (maglev) the system could be out of sight and out of mind for whatever benefits that may be for financial and aesthetic appeal. Considering that there are continental trenches around continents the choice of location worldwide is almost infinitely variable. While it is true that larger facilities have lower costs there may be merit in multiple facilities for the various multiple countries that will be involved.
- Population density in the area being mined has positive and negative benefits. Lots of people help construction and product use. Zero people help minimize resistance to the project. Considering the scale of the project zero people would seem preferred and the Western Sahara would be preferred on land or under-seas for hydraulic operation.

Conservation of the Environment: Unless the entire world's continental drift is totally constrained the creation of mountains, volcanoes, rain, mud, sand, rocks etc., will be conserved over time. Mass is neither being created nor destroyed, rather some of it is being returned to where it was at another time. Basically material on the continental shelf is being returned to the continental trenches. What people do with the product power will either help or harm the environment. The creation of the power provides the opportunity to mankind to decide the environments' and their own fate.

What Scale is needed? For the entire United States requirement in 2006 of 4.5*10^12 kwh or 600 gigawatts generating capacity the facility would be 300 times the scale of three gorges project in China or the U.S. Grand Cooley dam project using similar water flow and head. With a 50 fold head advantage the scale would be 300/50= 6 times that of either of these facilities. Assuming a water flow of 3 fold increase for that facility would require an overall 2 fold increase over either of these facilities. The total world electric demand would be satisfied by 10 such facilities.

Where are the facilities located? Considering that there are continental trenches at the boundary of the continents and over millions of years of erosion by wind and water, that have accumulated on continental shelves, they can be almost any where one desires. At some locations like Chile the continental trench is close by (e.g. 60 miles to 20,000 ft deep water) as well as having high elevation mountains near the coast. Others have broader continental shelves. It is not unreasonable to consider the exits (mouth) of major large rivers but the continental trenches may be less deep and farther away. This article is not a geological survey of the most economic locations; rather it is to observe that there are a very large number of candidates.

When will these facilities be built? Probably the time frame is 100 years in the future, say 2120. Plants are seldom built to provide greater than 20 years of visualizable needed capacity so something like 5 generation of concept and engineering evaluation followed by plant expansion will likely occur.

How much will electricity cost from a plant that is 300 greater capacity than the at present world's largest? If $0.10 will buy a kilowatt hour today then in 2012 dollars it will cost $0.10/(300)^(1/2)=$0.006/kwh.

Why will Continental Drift Power (CDP) be preferred? Depletion of oil and gas reserves are forecast to diminish significantly in 100 years, but it will be scale of operation and the low cost associated with the largest single plant output that drives Continental Drift Power (CDP) over the options of thousands of oil/gas wells and thousands of miles of distribution piping for its delivery (basically a 1 dimension activity). Green energy is basically a one year at a time event (in 2 dimensions), yet the proposed continental drift energy source has been accumulating for millions of years. In a way it has a 10^6 scale advantage over any earthbound green energy (and is scalable in 3 dimensions). Energy sources that have advantages of scale and accumulation over time include continental shelves, coal (strip mining is essentially 2 dimensional), oil, gas and nuclear energy materials, and are concentrated forms of energy. The competition between these accumulated sources will continue to play out on a cost basis.

How will Continental Drift Power (CDP) come about. I suppose it is not fair to misappropriate Winston Churchill and say "*It is not the end,* nor even the beginning of the end, but it is the end of the beginning" but the long road ahead would seem to make it appropriate. The need for energy from something other than fossil fuels is well known but the scale of presently discussed alternatives are so low as to make them expensive and ineffective in meeting the needs of the future. In the end it will be the scale, and accumulation of energy that will make continental drift energy prevail.

Is CDP conserved and sustainable? I am reminded of deriving differential equations in college wherein:*Input – Output – Reaction=Accumulation.* The input output difference is our desired power with zero chemical reaction required and the accumulation is over millions and millions of years. Over similar time frames it is conserving and fun to rewrite the equation: *Input – Output – Reaction –Accumulation = 0.* In other words it is conserving and sustainable for the human race for some time to come.

Is there Historical president or perspective? Tearing down mountains and dumping them in the sea is not a new concept but doing so to generate power for the world may be. There is a centuries old fable (7) in China where mountains are moved for the convenience of the human residents. Using hoes and baskets the family set out to dig up the mountain and dump it in the sea. Persisting for generation after generation ad infinitum they expected to accomplish their task; the moral of the story being many tasks appear unachievable but with determination anything can be achieved. *It has also been exemplified in Mao Tse Tung's telling of "Yu Gong Yi Shan," in a very famous speech.*

No philosophical or political motive for this article is intended though it is agreed that it will be the people of the world that will cause the mountains to be moved to generate the power. It is, after all, people that build the facilities in 3 dimensions over time and consume the product that is produced.

Rather it is to emphasize that the Scale of this proposal is markedly greater than the fables (hoes and baskets, vs. hydroelectric dams and volumetric flows greater than major rivers) and that the accumulated material (ground up mountain over millions of years vs. the original solid mountain) provides the energy basis for and from moving mountains.

Vertical Farming:

The use of electricity for light source for farming in a vertical as well as areal dimension (2 horizontal) is not new and has been discussed in the Economist (9). 'Vertical farming: does it really stack up". The conclusion there is correct, namely that the economics of doing so demand large scale low cost electricity. Given large scale low cost electricity farming can become 3 dimensional and therefore can be expanded at a greater rate than the growth rate of the population. In the very long term feeding the population can become a limitation that can be avoided by making farming a 3 dimensional activity.

World Population:

Presumably the world energy needs will grow with its population. Regardless of whose estimate one uses a significant increase beyond present will be required (10). Production of energy will preferably be proportional to the population, or greater, simply emphasizing the need for very large power generating facilities. Development of continental drift power (CDP) can presumably meet the demand within the next 100 years where population may increase 4 fold and continental drift power (CDP) can increase by 300 fold for each facility and 3000 fold for the world. Enough margin for growth and reduced costs too.

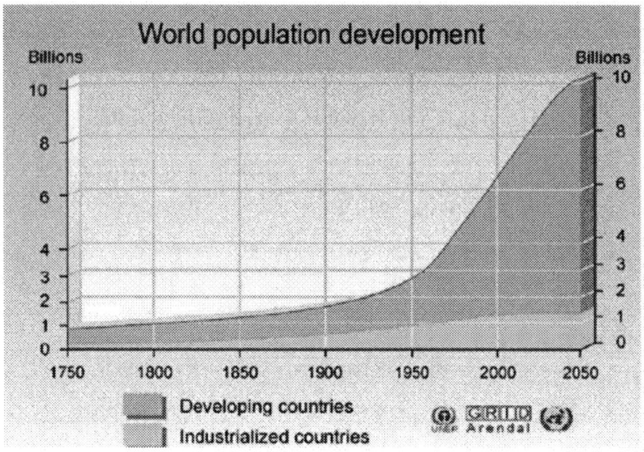

Will there be adequate Financial Capability: The financial capacity to do theses large projects will grow in time with the population hence should not be a limitation. For example the U. S. GDP alone may grow 10^4 over the 100 year period.

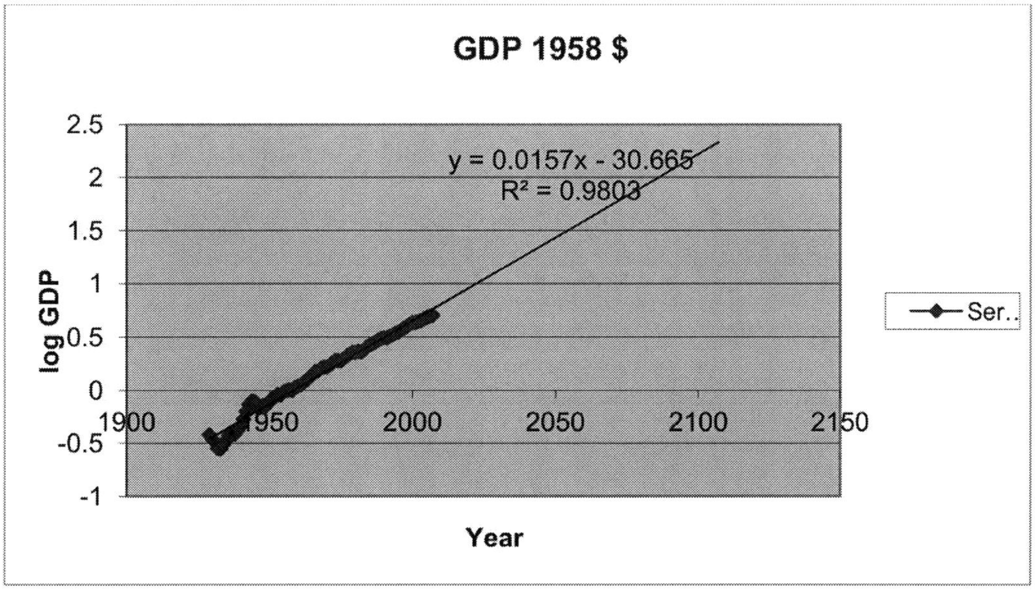

Other Benefits: All of the results (benefits) listed on page 2 naturally flow from very large scale availability of electric power at low cost. Nearly all are unattainable absent low cost power, e.g. the hydrogen economy, no green house gases, electric cars and airplanes, waste disposal reduced to total recycle, atmospheric modifications, vertical farming, even unlimited growth of the human race. And the visualizable list goes on limited only by ones imagination and can include space exploration, human colonization of planets and stars for the propagation of mankind etc. , etc.

Summary:

(1) The combined accumulated energy sources that make power from continental drift possible are:

- The radioactive decay of elements in the earth's interior that cause continental drift.
- The sun powered wind that erodes the mountains, evaporates the seas and causes rain to fall.
- The potential energy from elevated mass acted upon by gravity.
- Subduction Zones. When two oceanic plates collide, the younger of the two plates, because it is less dense, will ride over the edge of the older plate.
- Continental drift power (CDP) has tremendous scale, ergo cost, advantages vs. other alternative power sources.
- Continental drift power (CDP) is conservative and sustainable.

(2) A velocity enhancement over continental drift of 10^10 is achieved by gravitationally transporting mass to trenches at a higher velocity than the continental drift rate.

(3) The low cost of Continental Drift Power (CDP) owing to its scale will become the preferred alternative power source.

(4) "*It is not the end*, nor even the beginning of the end, but it is the end of the beginning", per Churchill.

References:

1. *http://scienceworld.wolfram.com/biography/Archimedes.html*
2. This article: Copyright: Donald I. Garnett, 14206 San Felipe Dr., Corpus Christi, Texas, Tuesday, 26 April 2011.
3. *http://experiencecurves.com/ecblog/*
4. *http://library.thinkquest.org/17457/platetectonics/2.php*
5. *http://www.stockton.edu/~hozikm/geol/Courses/The%20Earth/Content%20Web%20Pages/Bugielski/webpage.htm*
6. *http://www.eia.gov/emeu/iea/Notes%20for%20Table%206_2.html*
7. *http://english.cri.cn/4426/2007/01/15/167@185195.htm*
8. *http://www.story-lovers.com/listschinesestories.html*
9. *http://www.economist.com/node/17647627*
10. http://www.sustainablescale.org/areasofconcern/population/populationandscale/quickfacts.aspx
11. *http://physics-edu.org/energy_from_continental_drift.htm*

AUTHOR INDEX